T0178493

Springer INdAM Series

Volume 26

More information about this series at http://www.springer.com/series/10283

Filippo Bracci
Editor

Geometric Function Theory
in Higher Dimension

 Springer

Editor
Filippo Bracci
Dipartimento di Matematica
Università di Roma Tor Vergata
Rome, Italy

ISSN 2281-518X   ISSN 2281-5198   (electronic)
Springer INdAM Series
ISBN 978-3-030-10319-4   ISBN 978-3-319-73126-1   (eBook)
https://doi.org/10.1007/978-3-319-73126-1

Preface

This book contains a collection of papers written by participants of the INDAM Workshop "Geometric Function Theory in Higher Dimension", held at the Palazzone in Cortona, Italy from September 5 to 9, 2016.

The meeting brought together international experts on the subject of geometric function theory in one and several variables, as well as young researchers, in order to discuss old and new open problems in the area and stimulate new joint work. The talks were mainly focusing on open problems and recent achievements on the subject. In the same spirit, we asked the authors of this volume to write their contributions focusing on recent developments in the sector and challenging new problems.

Geometric function theory in one dimension has been and still is one of the main streams of research in mathematics since early in the last century. One of the main tools is the deterministic Loewner differential equation, which was introduced by Charles Loewner in 1923 for the study of problems in complex analysis. This is the prototype of a mathematical framework which was initially developed for entirely intrinsic reasons long before the appearance of applications, among one may mention the discovery by Oded Schramm in 1999 of a *stochastic* version of Loewner's equation (SLE) relating complex analysis and probability in a completely novel way. In the meantime, Loewner's equation and some techniques from the one dimensional case have been extended to several complex variables. A number of open problems have now been solved and several new lines of investigation have evolved. Nevertheless, as has been apparent from the outset, some questions in Loewner theory in higher dimensions lead beyond the classical theory, and new tools are needed. For instance, Fatou-Bieberbach phenomena make things much more subtle and new techniques are being developed to handle such problems.

Viewed abstractly, Loewner's theory deals with the interplay of three main objects: a non-autonomous holomorphic vector field which is semicomplete for almost every fixed nonnegative time—the so-called *Herglotz vector field*, a family of time depending image growing univalent mappings—a *Loewner chain*, and a bi-parametric family of univalent self-maps satisfying a semigroup-type condition— an *evolution family*. The theory developed so far asserts that there is a one-to-one correspondence among these objects, which is given by the Loewner ODE and the Loewner-Kufarev PDE. It is then clear that geometric properties of the images of a Loewner chain correspond to analytic properties of the associated Herglotz vector field which in turn are related to the dynamical properties of the associated evolution family.

In order to properly understand the analytical properties of Herglotz vector fields one is forced to study the so-called infinitesimal generators, namely, those autonomous holomorphic vector fields which are semicomplete, i.e., those whose flow extends for all nonnegative time. In this framework, the main problems are related to the boundary behavior of infinitesimal generators defined on domains in the multidimensional complex space. On the other hand, the dynamics of evolution families can be understood via iteration theory (and boundary behavior) and linearization (or more generally the study of the so-called linear fractional models) of holomorphic self-maps and semigroups of a given domain.

Finally, the geometry of the image of univalent mappings, which is, strictly speaking the content of geometric function theory, can be studied with different methods. As already remarked, the presence of new typical phenomena of higher dimension, such as the lack of a uniformization theorem, the Fatou-Bieberbach phenomena, Runge-ness problems, and resonances, makes things much more complicated and demands the development of new techniques (especially $\bar{\partial}$-equations, complex Monge-Ampère equations, approximation techniques, Kobayashi's metric). Nonetheless, in the case of the Euclidean ball, the class S^0 of maps which admit the so-called parametric representation (i.e., that can be embedded into a normal Loewner chain), has been studied with success. Such a class, which is compact in the Fréchet space of holomorphic mappings from the ball, is the analogue of the class of normalized univalent function in the unit disc. Growth estimates, radii problems and characterization of univalent functions in S^0 have been found. In higher dimensions, this class is a proper subclass of the class $S(\mathbb{B}^n)$ consisting of normalized univalent mappings of the ball $\mathbb{B}^n$, but, in contrast to the one-dimensional case, in higher dimensions there exist mappings in $S(\mathbb{B}^n)$ which cannot be embedded in normal Loewner chains. In fact, the class $S(\mathbb{B}^n)$ is not compact in dimension $n \geq 2$. Although at a first sight the class S^0 seems to be too "small" to give information properly on every normalized univalent map of the ball, in fact, every univalent map which can be embedded into a normalized Loewner chain, sits in S^0 up to post-composition with an entire univalent map. Also, in strict contrast to the one-dimensional case, and quite surprisingly, the (restrictions to the ball of) automorphisms of the space are dense in S^0 and there exist support points in S^0—namely, maps that maximize some linear bounded operator—which are bounded.

The contributions to this volume consider the aforementioned setting from different points of views, in one or several variables, and provide new insights into the full theory, explain relevant conjectures in the area, or address interesting open problems.

The INDAM Workshop was organized by myself, Ian Graham, Gabriela Kohr, Oliver Roth, and David Shoikhet. Neither the workshop itself nor this volume would have been possible without the help and contributions of Ian, Gabi, Oliver and David. To all of them, my deepest gratitude and thanks.

Rome, Italy   Filippo Bracci
October 27, 2017

Contents

About the Editor

Prof. Filippo Bracci obtained his Ph.D. in Mathematics at the University of Padova in 2001. He is currently full professor at Università di Roma "Tor Vergata". He was the principal investigator of the ERC project "HEVO". His research interests include complex analysis, several complex variables, and holomorphic dynamics.

The Embedding Conjecture and the Approximation Conjecture in Higher Dimension

Matteo Fiacchi

Abstract In this paper we show the equivalence among three conjectures (and related open questions), namely, the embedding of univalent maps of the unit ball into Loewner chains, the approximation of univalent maps with entire univalent maps and the immersion of domain biholomorphic to the ball in a Runge way into Fatou-Bieberbach domains.

Keywords Loewner theory · Embedding problems · Approximation of univalent maps

Mathematics Subject Classification 32E30, 32A10

1 Preliminaries

Let $\mathbb{B}^n := \{z \in \mathbb{C}^n : ||z|| < 1\}$ be the unit ball of $\mathbb{C}^n$ and

$$\mathcal{S} := \{f : \mathbb{B}^n \longrightarrow \mathbb{C}^n \text{ univalent s.t. } f(0) = 0, (df)_0 = \mathsf{Id}\}.$$

A *(normalized) Loewner chain* is a continuous family of univalent mappings $(f_t : \mathbb{B}^n \longrightarrow \mathbb{C}^n)_{t \geq 0}$ with $\Omega_s := f_s(\mathbb{B}^n) \subseteq f_t(\mathbb{B}^n)$ for $s \leq t$ and such that $f_t(0) = 0$ and $(df)_0 = e^t \mathsf{Id}$ for all $t \geq 0$. We also define the *Loewner range* of a normalized Loewner chain as

$$R(f_t) = \bigcup_{t \geq 0} \Omega_t \subseteq \mathbb{C}^n.$$

M. Fiacchi (✉)
Dipartimento Di Matematica, Università di Roma "Tor Vergata", Roma, Italy
e-mail: fiacchi@mat.uniroma2.it

© Springer International Publishing AG, part of Springer Nature 2017
F. Bracci (ed.), *Geometric Function Theory in Higher Dimension*,
Springer INdAM Series 26, https://doi.org/10.1007/978-3-319-73126-1_1

The Loewner range is always biholomorphic to $\mathbb{C}^n$, although, for $n > 1$, it can be strictly contain in $\mathbb{C}^n$ (see [2]). Furthermore a normalized Loewner chain $(f_t)_{t \geq 0}$ is *normal* if $\{e^{-t} f_t(\cdot)\}_{t \geq 0}$ is a normal family.

We say that $f \in \mathcal{S}$ *embeds into a Loewner chain* $(f_t)_{t \geq 0}$ if $f_0 = f$. Then we can define

$$\mathcal{S}^1 := \{f \in \mathcal{S} : f \text{ embeds into a normalized Loewner chain}\}$$

and, following ideas of I. Graham, H. Hamada and G. Khor in [6] and [7], we can define the class

$$\mathcal{S}^0 := \{f \in \mathcal{S} : f \text{ embeds into a normal Loewner chain}\}.$$

The class $\mathcal{S}^0$ is compact [5]. We also denote with

$$\mathcal{S}(\mathbb{C}^n) := \{\Psi : \mathbb{C}^n \longrightarrow \mathbb{C}^n \text{ univalent} : \Psi(0) = 0, (d\Psi)_0 = \mathsf{Id}\}$$

the space of normalized entire univalent functions. We recall also the following result

Proposition 1.1 ([5]) *If $(f_t)_{t \geq 0}$ is a normalized Loewner chain, then there exist an unique normalized biholomorphism $\Psi : \mathbb{C}^n \longrightarrow R(f_t)$ and normal Loewner chain $(g_t)_{t \geq 0}$ such that for each $t \geq 0$*

$$f_t = \Psi \circ g_t.$$

In particular in the case $t = 0$, we have that if $f \in \mathcal{S}^1$ then there exist $\Psi \in \mathcal{S}(\mathbb{C}^n)$ and $g \in \mathcal{S}^0$ such that $f = \Psi \circ g$. In particular, we have the following decomposition

$$\mathcal{S}^1 = \mathcal{S}(\mathbb{C}^n) \circ \mathcal{S}^0.$$

2 Embedding Problems in Loewner Theory

In one dimension all normalized univalent mapping on the disk embed into a normal Loewner chain [8] and all normalized Loewner chains are, in fact, normal. Namely, for $n = 1$,

$$\mathcal{S}^0 = \mathcal{S}^1 = \mathcal{S}.$$

On the other hand, in higher dimension the situation becomes much more complicated. A natural question in Loewner theory in several complex variables, coming from the parallel with dimension one, is the following:

does every normalized univalent map on the ball embed into a normalized Loewner chain?

The class $\mathcal{S}^0$ is compact whereas $\mathcal{S}$ and $\mathcal{S}^1$ are not [5, Chapter 8]. Hence, there are some normalized univalent functions on the ball that do not embed into a normal Loewner chain (i.e. $\mathcal{S}^0 \subsetneq \mathcal{S}$). But, is it possible to embed them into a normalized Loewner chain? The question is still open, and we have the following

Conjecture 2.1 For $n \geq 2$ we have

$$\mathcal{S} = \mathcal{S}^1.$$

By Proposition 1.1 this is equivalent to

$$\mathcal{S} = \mathcal{S}(\mathbb{C}^n) \circ \mathcal{S}^0.$$

This conjecture explains very effectively the differences of $\mathcal{S}$ between one and several variables, because, for $n = 1$, $\mathcal{S}(\mathbb{C}) = \{\mathsf{Id}_\mathbb{C}\}$.

Remark 2.2 Let $\Psi \in \mathcal{S}(\mathbb{C}^n)$. Then $\Psi|_{\mathbb{B}^n} \in \mathcal{S}^1$. Indeed, Ψ can be embedded into the normalized Loewner chain $\{\Psi(e^t z)\}$.

A striking difference between one and several complex variables concern entire univalent mappings: in one dimension, entire univalent maps are only affine transformations while in several variables they are plenty of entire univalent maps which are not affine transformations and not even surjective. This allows for approximation's results which are not possible in dimension 1. A classical result in this sense is the following

Theorem 2.3 (Andérsen-Lempert [1]) *For $n \geq 2$, let $D \subseteq \mathbb{C}^n$ be a starlike domain and let $f : D \longrightarrow \mathbb{C}^n$ be an univalent mapping. Then $f(D)$ is Runge if and only if f can be approximated uniformly on compacta of D by automorphisms of $\mathbb{C}^n$, i.e. for every compact subset K of D, $\epsilon > 0$ there exists an automorphism Φ of $\mathbb{C}^n$ such that $\max_{z \in K} ||f(z) - \Phi(z)|| < \epsilon$.*

It is natural to wonder if a similar result is valid for no-Runge mappings, then we have the following

Conjecture 2.4 (Generalized Anderson-Lempert Theorem [GAL]) For $n \geq 2$, let $f \in \mathcal{S}$ with $f(\mathbb{B}^n)$ not Runge. Then there exists a sequence of entire univalent maps which approximate f uniformly on compacta of $\mathbb{B}^n$.

Due to Theorem 2.3, it is clear that if Conjecture 2.4 holds and f has no Runge image, then entire maps which approximate f uniformly on compacta of $\mathbb{B}^n$ cannot have Runge image. Their images are thus *Fatou-Bieberbach domains* in $\mathbb{C}^n$ which are not Runge in $\mathbb{C}^n$.

Remark 2.5 In the class $\mathcal{S}^1$, [GAL] holds. Indeed, let $f \in \mathcal{S}^1$, then by Proposition 1.1 there exists $\Psi \in \mathcal{S}(\mathbb{C}^n)$ and $g \in \mathcal{S}^0$ such that $f = \Psi \circ g$. Now, $g(\mathbb{B}^n)$ is Runge, then by the Andérsen-Lempert theorem there exists a sequence of normalized automorphisms $\{\Phi_k\}_{k\in\mathbb{N}}$ that converges to g uniformly on compacta of $\mathbb{B}^n$. Then, also $\{\Psi \circ \Phi_k\}_{k\in\mathbb{N}} \in \mathcal{S}(\mathbb{C}^n)$ converges to $f = \Psi \circ g$ i.e. [GAL] holds in $\mathcal{S}^1$.

This conjecture is equivalent to the density of $\mathcal{S}^1$ in $\mathcal{S}$. Indeed

Proposition 2.6 *For $n \geq 2$, the following are equivalent*

(1) [GAL] holds;
(2) $\overline{\mathcal{S}^1} = \mathcal{S}$ in the topology of the uniform convergence on compacta.

Proof (1) $\Rightarrow$ (2) Let $f \in \mathcal{S}$. Then there exists a sequence $\{\Psi_k\} \subset \mathcal{S}(\mathbb{C}^n)$ such that $\Psi_k \longrightarrow f$ uniformly on compacta of $\mathbb{B}^n$ (this follows by Andérsen-Lempert Theorem if f is Runge or by (1) otherwise). But $\Psi_{k|\mathbb{B}^n} \in \mathcal{S}^1$ by Remark 2.2, hence (2) holds.

(2) $\Rightarrow$ (1) By Remark 2.5 the set

$$\mathcal{A} := \{f \in \mathcal{S} : f = \Psi_{|\mathbb{B}^n} \text{ for some } \Psi \in \mathcal{S}(\mathbb{C}^n)\}$$

is dense in $\mathcal{S}^1$ i.e. $\mathcal{S}^1 \subseteq \overline{\mathcal{A}}$. Therefore by (2)

$$\mathcal{S} = \overline{\mathcal{S}^1} \subseteq \overline{\mathcal{A}} \subseteq \mathcal{S} \quad \Rightarrow \quad \overline{\mathcal{A}} = \mathcal{S}.$$

This clearly implies [GAL]. □

We denote

$$\mathcal{S}_R := \{f \in \mathcal{S} : f(\mathbb{B}^n) \text{ is Runge}\}.$$

In order to find conditions for embedding of univalent functions, a lot of results have been proved during these years. Some of these concern the regularity of the boundary of $f(\mathbb{B}^n)$. Arosio, Bracci and Wold in 2013 in [2] proved the following

Theorem 2.7 *Let $n \geq 2$ and $f \in \mathcal{S}_R$. If $\Omega := f(\mathbb{B}^n)$ is a bounded strongly pseudoconvex domain with $\mathcal{C}^\infty$ boundary and $\overline{\Omega}$ is polynomially convex, then $f \in \mathcal{S}^1$.*

A simple corollary of this theorem is that all functions in $\mathcal{S}_R$ that extend in a univalent and Runge way to a neighborhood of $\overline{\mathbb{B}^n}$ embed into a normalized Loewner chain.

Corollary 2.8 *Let $n \geq 2$ and $f \in \mathcal{S}_R$. If there exist $r > 1$ and $F \in \mathcal{S}_R(r\mathbb{B}^n)$ s.t. $F_{|\mathbb{B}^n} = f$, then $f \in \mathcal{S}^1$.*

Proof We use Theorem 2.7. Obviously $\Omega := f(\mathbb{B}^n)$ has $\mathcal{C}^\infty$ boundary and Ω is a strongly pseudoconvex domain (because $\mathbb{B}^n$ is and since the biholomorphism is defined in a neighborhood of $\overline{\mathbb{B}^n}$ preserves the strongly pseudoconvexity). Now $\forall \, s \in (0, r) \; \Omega_s := F(s\mathbb{B}^n)$ are Runge, thus $\overline{\Omega}_s$ has a fundamental system of open neighborhoods which are pseudoconvex and Runge, hence $\overline{\Omega}_s$ is polynomially convex. □

Following the ideas contained in these results, a natural question is the following:
does every normalized univalent function of the ball that extends in a univalent way to a neighborhood of the closed ball, embeds into a normalized Loewner chain (without requiring that $f \in S_R$)?

Therefore we have the following

Conjecture 2.9 ([EXT]) *Let $n \geq 2$ and $f \in S$. If exist $r > 1$ and $F \in S(r\mathbb{B}^n)$ s.t. $F_{|\mathbb{B}^n} = f$, then $f \in S^1$.*

Remark 2.10 The conjecture [EXT] is equivalent to requiring that for every $f \in S$ and $r \in (0, 1)$ the mappings $f_r(z) := \frac{1}{r}f(rz) \in S$ embed into a normalized Loewner chain.

Also this conjecture turns out to be equivalent to [GAL].

Proposition 2.11 *For $n \geq 2$, the following are equivalent*

(1) [GAL] holds (equivalently $\overline{S^1} = S$);
(2) [EXT] holds.

Proof (1) $\Rightarrow$ (2) Given $f \in S$ such that there exist $r > 1$ and $F \in S(r\mathbb{B}^n)$ with $F_{|\mathbb{B}^n} = f$, by [GAL] F can be approximated by $\Psi_k \in S(\mathbb{C}^n)$ uniformly on compacta of $\mathbb{B}^n$. Now there exists $k_0 > 0$ such that for each $k > k_0$ we have

$$f(\overline{\mathbb{B}^n}) =: \overline{\Omega} \subset \Psi_k(\mathbb{B}^n)$$

and $\Psi_k^{-1} \longrightarrow F^{-1}$ uniformly on $\overline{\Omega}$. But $F^{-1}(\overline{\Omega}) = \overline{\mathbb{B}^n}$ is a strongly convex set and since strong convexity is an open condition (in the C^2 topology), there exist $k_1 > k_0$ and $\Psi := \Psi_{k_1}$ such that $\Psi^{-1}(\overline{\Omega})$ is strongly convex. Therefore f can be embedded into the Loewner chain

$$f_t(z) := \Psi(e^t \cdot \Psi^{-1}(f(z))).$$

(2) $\Rightarrow$ (1) Given $\{r_k\}_{k \in \mathbb{N}} \subset (0, 1)$ a sequence that converges to 1, for each $f \in S$ we can define $f_k(z) := \frac{1}{r_k}f(r_kz) \in S$. Obviously $f_k \longrightarrow f$ uniformly on compacta of $\mathbb{B}^n$, and $f_k \in S^1$ for each $k \in \mathbb{N}$ by [EXT]. Hence $\overline{S^1} = S$. ☐

We recall the following result, that easily descends from the Docquier-Grauert Theorem [3].

Proposition 2.12 *Let $(f_t)_{t \geq 0}$ be a normalized Loewner chain. Then for each $0 < s \leq t$ the couple $(f_s(\mathbb{B}^n), f_t(\mathbb{B}^n))$ is Runge. Therefore for each $t \geq 0$ also the couple $(f_t(\mathbb{B}^n), R(f_t))$ is Runge.*

Given $(f_t)_{t \geq 0}$ a normalized Loewner chain, by the previous proposition we have that $f_0(\mathbb{B}^n)$ is Runge into its Loewner range $R(f_t)$ that is biholomorphic to $\mathbb{C}^n$. Thus a necessary condition in order to obtain embedding of all the univalent functions into a Loewner chain is that all the open sets of $\mathbb{C}^n$ biholomorphic to $\mathbb{B}^n$ have to be Runge in $\mathbb{C}^n$ or in some Fatou-Bieberbach domain. Thus we have the following

Conjecture 2.13 ([FBR]) Let $D \subseteq \mathbb{C}^n$ a domain biholomorphic to the ball $\mathbb{B}^n$, which is not Runge in $\mathbb{C}^n$. Then there exists $\Omega \subseteq \mathbb{C}^n$ a Fatou-Bieberbach domain with $D \subseteq \Omega$ such that (D, Ω) is a Runge pair.

Related to the previous conjecture, we have the following strong version of [GAL].

Conjecture 2.14 (Strong Generalized Andérsen-Lempert Theorem [GAL^s]) For $n \geq 2$, let $f \in \mathcal{S}$ with $f(\mathbb{B}^n)$ not Runge. Then there exist a Fatou-Bieberbach domain Ω and a sequence of univalent mappings $\Psi_k \in \mathcal{S}(\mathbb{C}^n)$ with $\Psi_k(\mathbb{C}^n) = \Omega$ for each k, which converges to f uniformly on compacta of $\mathbb{B}^n$.

We have also the following weaker formulations

Conjecture 2.15 ([FBR$_a$]) Let $f \in \mathcal{S}$, then for each $r \in (0, 1)$ exists $\Omega_r \subseteq \mathbb{C}^n$ a domain biholomorphic to $\mathbb{C}^n$ with $f(r\mathbb{B}^n) \subseteq \Omega_r$ such that $(f(r\mathbb{B}^n), \Omega_r)$ is a Runge pair.

Conjecture 2.16 ([GAL$_a^s$]) Let $f \in \mathcal{S}$ with $f(\mathbb{B}^n)$ not Runge, then for each $r \in (0, 1)$ there exist a Fatou-Bieberbach domain Ω_r and a sequence of univalent mappings $\Psi_k^{(r)} \in \mathcal{S}(\mathbb{C}^n)$ with $\Psi_k^{(r)}(\mathbb{C}^n) = \Omega_r$ for each k, which converges to f_r uniformly on compacta of $\mathbb{B}^n$.

Obviously [FBR] and [GAL^s] imply respectively [FBR$_a$] and [GAL$_a^s$], and furthermore they are equivalent two by two.

Proposition 2.17 *For $n \geq 2$, we have*

(1) [FBR] and [GAL^s] are equivalent.
(2) [FBR$_a$] and [GAL$_a^s$] are equivalent.

Proof The two proofs are essentially the same, so we only prove (1).

([FBR]$\Rightarrow$[GAL^s]) Let $f \in \mathcal{S}$ and $\Psi \in \mathcal{S}(\mathbb{C}^n)$ be a Fatou Bieberbach mapping such that $(f(\mathbb{B}^n), \Psi(\mathbb{C}^n))$ is Runge. Now since Runge-ness is a property invariant under biholomorphism, $((\Psi^{-1} \circ f)(\mathbb{B}^n), \mathbb{C}^n)$ is Runge. By the Andérsen-Lempert theorem $\Psi^{-1} \circ f$ is approximable by automorphisms Φ_k, then $\Psi \circ \Phi_k \in \mathcal{S}(\mathbb{C}^n)$ converges to f, i.e. [GAL^s] holds.

([GAL^s]$\Rightarrow$[FBR]) Let $f \in \mathcal{S}$ and $\{\Psi_k\}_{k \in \mathbb{N}} \subset \mathcal{S}(\mathbb{C}^n)$ a sequence of normalized entire univalent mappings with the same image Ω that approximate f. Now $f(\mathbb{B}^n) \subseteq \Omega$, then the sequence $\{\Psi_0^{-1} \circ \Psi_k\}_{k \in \mathbb{N}}$ converges to $\Psi_0^{-1} \circ f$. Note that $\Psi_0^{-1} \circ \Psi_k$ are normalized automorphisms for each $k \in \mathbb{N}$, then by the necessity condition of Andérsen-Lempert theorem $\Psi_0^{-1} \circ f$ has to be Runge. Finally, by invariance of Runge-ness under biholomorphisms, $(f(\mathbb{B}^n), \Psi_0(\mathbb{C}^n) = \Omega)$ is Runge. $\square$

Proposition 2.18 *For $n \geq 2$, we have*

(1) $[GAL_a^s]$ implies $[GAL]$.
(2) $[EXT]$ implies $[FBR_a]$.

Moreover, by Propositions 2.11 and 2.17 $[FBR_a]$, $[GAL]$, $[GAL_a^s]$ and $[EXT]$ conjectures are equivalent to $\overline{S^1} = S$.

Proof ($[GAL_a^s] \Rightarrow [GAL]$) Let $f \in S$ with $f(\mathbb{B}^n)$ non Runge and $K \subseteq \mathbb{B}^n$ a compact. Then there exists $r \in (0,1)$ such that $K \subseteq r\mathbb{B}^n$. Thanks to $[GAL_a^s]$, there exists a normalized Fatou Bieberbach mapping $\Psi \in S(\mathbb{C}^n)$ such that

$$\max_{z \in \frac{1}{r}K} ||f_r(z) - \Psi(z)|| < \epsilon/r$$

then

$$\max_{z \in K} ||f(z) - r\Psi(z/r)|| = r\max_{z \in \frac{1}{r}K} ||f(rz)/r - \Psi(z)|| < \epsilon$$

i.e. $[GAL]$ holds.
($[EXT] \Rightarrow [FBR_a]$) Fix $f \in S$. For each $r \in (0,1)$ by $[EXT]$ f_r embeds into a normalized Loewner chain, then $f_r(\mathbb{B}^n)$ is Runge in its Loewner range. □

To summarize, we have the following scheme

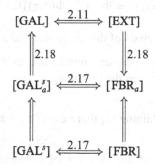

Remark 2.19

(1) S_R is close and by the Andérsen-Lempert theorem $S_R \subseteq \overline{S^1}$. Therefore, if S^1 were closed then $S_R \subseteq S^1$ i.e. every univalent function with Runge image would be embedded into a normalized Loewner chain.
(2) If $[FBR]$ holds and $S_R \subseteq S^1$, then every normalized univalent function would be embedded into a normalized Loewner chain. Indeed if $f \in S$ with $f(\mathbb{B}^n)$ not Runge, by $[FBR]$ there exists $\Psi \in S(\mathbb{C}^n)$ such that $(\Psi^{-1} \circ f)(\mathbb{B}^n)$ is Runge, hence, if $S_R \subseteq S^1$ $\Psi^{-1} \circ f \in S^1$. Then $f \in S^1$.

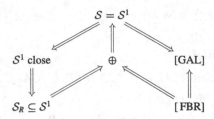

3   Entire-Convexshapelike Domains

In this section we give a proof of the following theorem, that is the generalization of Theorem 2.7 under the additioned hypothesis that [GAL] holds.

Theorem 3.1 *Let $n \geq 2$ and suppose [GAL] holds. If $f \in S$ has the property that $f(\mathbb{B}^n)$ is a bounded strongly pseudoconvex domain with C^∞ boundary then $f \in S^1$.*

In [2], in order to prove Theorem 2.7 it was introduced the concept of *convexshapelike domains*: $\Omega \subseteq \mathbb{C}^n$ is convexshapelike if there exists Φ an automorphism of $\mathbb{C}^n$ such that $\Phi(\Omega)$ is convex in $\mathbb{C}^n$. This kind of domains are very useful in the study of embedding problems, indeed given $f \in S$ if $f(\mathbb{B}^n)$ is convexshapelike then f embeds into the Loewner chain

$$f_t(z) := \Phi^{-1}(e^t \cdot \Phi(f(z))).$$

According to [2], a Loewner chain of this form is called a *filtering Loewner chain*.

Definition 3.2 A normalized Loewner chain $(f_t)_{t \geq 0}$ in $\mathbb{B}^n$ is a *filtering normalized Loewner chain* if $\Omega_t := f_t(\mathbb{B}^n)$ has the following properties:

(1) $\bar{\Omega}_s \subseteq \Omega_t$ for each $0 \leq s < t$;
(2) for each open set U containing Ω_s there exists $t_0 > s$ such that $\Omega_t \subset U$ for all $t \in (s, t_0)$.

A natural generalization of the concept of convexshapelike domains is the following

Definition 3.3 $\Omega \subseteq \mathbb{C}^n$ is *entire-convexshapelike domains* if there exists $\Psi : \mathbb{C}^n \longrightarrow \mathbb{C}^n$ an entire univalent mapping with $\Omega \subseteq \Psi(\mathbb{C}^n)$ s.t. $\Psi^{-1}(\Omega)$ is convex.

As before, if $f \in S$ and $f(\mathbb{B}^n)$ is entire-convexshapelike then f embeds into the filtering normalized Loewner chain

$$f_t(z) := \Psi(e^t \cdot \Psi^{-1}(f(z))),$$

where $\Psi : \mathbb{C}^n \longrightarrow \mathbb{C}^n$ is a normalized entire univalent mapping such that $\Psi^{-1}(f(\mathbb{B}^n))$ is convex.

Lemma 3.4 ([2]) *Let $\Omega \subseteq \mathbb{C}^n$ be a bounded strongly pseudoconvex domain with C^∞ boundary which is biholomorphic to $\mathbb{B}^n$. Then any $f \in C^2(\overline{\Omega}) \cap Hol(\Omega, \mathbb{C}^n)$ can be approximated uniformly on $\overline{\Omega}$ in C^2-norm, by functions in $Hol(\overline{\Omega}, \mathbb{C}^n)$.*

Therefore, in order to obtain Theorem 3.1, we prove the following

Proposition 3.5 *Let $n \geq 2$ and suppose [GAL] holds. If Ω is a bounded strongly pseudoconvex domain with C^∞ boundary, then Ω is entire-convexshapelike.*

Proof By Fefferman's Theorem [4] f extends to a diffeomorphism $f : \overline{\mathbb{B}^n} \longrightarrow \overline{\Omega}$. By Lemma 3.4 f^{-1} can be approximated in C^2 norms uniformly on $\overline{\Omega}$ by holomorphic maps defined on neighborhoods of $\overline{\Omega}$. Then there exists an open neighborhood U of $\overline{\Omega}$ and $h : U \longrightarrow \mathbb{C}^n$ an univalent mapping s.t. $h(\Omega)$ is a smooth strongly convex domain (because strongly convexity is an open condition). Without loss of generality, we can choose $h(U) = r\mathbb{B}^n$ with $r > 1$. Now, by [GAL], h^{-1} can be approximated uniformly on compacta of $h(U) = r\mathbb{B}^n$ by $\Psi_k \in \mathcal{S}(\mathbb{C}^n)$, then there exists $k_0 > 0$ such that for each $k > k_0$ we have

$$h^{-1}(\overline{\mathbb{B}^n}) = \overline{\Omega} \subset \Psi_k(\mathbb{B}^n)$$

and $\Psi_k^{-1} \longrightarrow h$ uniformly on $\overline{\Omega}$. But $h(\overline{\Omega}) = \overline{\mathbb{B}^n}$ is a strongly convex set, therefore there exist $k_1 > k_0$ and $\Psi := \Psi_{k_1}$ such that $\Psi^{-1}(\overline{\Omega})$ is strongly convex, i.e. Ω is entire-convexshapelike. $\qquad\square$

We recall a classical result of several complex variables.

Lemma 3.6 (Narasimhan) *Let $\Omega \subset \mathbb{C}^n$ be a domain with boundary C^2 in $p \in \partial\Omega$, and suppose that Ω is strongly Levi pseudoconvex in p. Then there exists an open neighborhood U of p and a biholomorphism $f : U \longrightarrow V \subset \mathbb{C}^n$ such that $f(U \cap \Omega)$ is a convex domain.*

Proposition 3.5 can be seen as a global version of Narasimhan's lemma (for domains biholomorphic to a ball and with C^∞ boundary).

We conclude with the following

Remark 3.7 Suppose the statement of Proposition 3.5 holds (i.e. every domain biholomorphic to a ball, with C^∞ boundary and strongly pseudoconvex, is entire-convexshapelike), then $\mathcal{S}^1$ is dense in $\mathcal{S}$. Indeed [EXT] holds: fix $f \in \mathcal{S}$, then for each $r \in (0, 1)$ $f_r(z) := \frac{1}{r}f(rz)$ is entire-convexshapelike and therefore it is in $\mathcal{S}^1$.

Acknowledgements The author would like to express his gratitude to Prof. Filippo Bracci for his availability and for introducing him into Loewner theory.

References

1. Andersen, E., Lempert, L.: On the group of holomorphic automorphisms of $\mathbb{C}^n$. Invent. Math. **110**, 371–388 (1992)
2. Arosio, L., Bracci, F., Fornaess Wold, E.: Embedding univalent functions in filtering Loewner chains in higher dimension. Proc. Am. Math. Soc. **143**, 1627–1634 (2015)

3. Docquier, F., Grauert, H.: Levisches problem und Rungescher Satz für Teilgebiete Steinscher Mannigfaltigkeiten. Math. Ann. **140**, 94–123 (1960)
4. Fefferman, C.: On the Bergman kernel and biholomorphic mappings of pseudoconvex domains. Invent. Math. **26**, 667–669 (1974)
5. Graham, I., Kohr, G.: Geometric Function Theory in One and Higher Dimensions. CRC Press, Boca Raton (2003)
6. Graham, I., Hamada, H., Kohr, G.: Parametric representation of univalent mappings in several complex variables. Can. J. Math. **54**, 324–351 (2002)
7. Kohr, G.: Using the method of Loewner chains to introduce some subclasses of biholomorphic mappings in $\mathbb{C}^n$. Rev. Roum. Math. Pures Appl. **46**, 743–760 (2001).
8. Pommerenke, C., Jensen, G.: Univalent functions. Vandenhoeck und Ruprecht, Göttingen (1975)

Fixed Points of Pseudo-Contractive Holomorphic Mappings

Mark Elin and David Shoikhet

Abstract We study conditions that ensure the existence of fixed points of pseudo-contractive mappings originally considered by Browder, Kato, Kirk and Morales. Specifically we consider holomorphic pseudo-contractions on the open unit ball of a complex Banach space which in general are not necessarily bounded. As a consequence, we obtain sufficient conditions for the existence and uniqueness of the common fixed point of a semigroup of holomorphic self-mappings and study its rate of convergence to this point.

Keywords Holomorphic maps · Fixed point theory · Contractive maps

Mathematics Subject Classification 37C25, 32A10

Let $\mathcal{D}$ be a domain (open connected subset) in a complex Banach space X. A mapping $F : \mathcal{D} \to X$ is said to be holomorphic if it is Frechét differentiable at each point $x \in \mathcal{D}$. By $\mathrm{Hol}(\mathcal{D}, X)$ we denote the set of all holomorphic mappings on $\mathcal{D}$.

In this paper we study conditions that ensure the existence of fixed points of F in $\mathcal{D}$. A standard situation in this study is when F is a holomorphic self-mapping of $\mathcal{D}$. In this case we write $F \in \mathrm{Hol}(\mathcal{D})$, where $\mathrm{Hol}(\mathcal{D})$ denotes the semigroup with respect to the composition operation of holomorphic self-mappings of $\mathcal{D}$.

In the one-dimensional case, the Grand Fixed Point Theorem due to Denjoy and Wolff (see, for example, [7, 25, 27, 29] and [30]) implies that if $\mathcal{D} = \Delta$ is the open unit disk in the complex plane $\mathbb{C}$, and $F \in \mathrm{Hol}(\Delta)$ has no fixed point on $\partial\Delta$, then F has a unique fixed point inside Δ.

M. Elin
Department of Mathematics, ORT Braude College, Karmiel, Israel
e-mail: mark_elin@braude.ac.il

D. Shoikhet (✉)
Department of Mathematics, Holon Institute of Technology, Holon, Israel
e-mail: davidsho@hit.ac.il

© Springer International Publishing AG, part of Springer Nature 2017
F. Bracci (ed.), *Geometric Function Theory in Higher Dimension*,
Springer INdAM Series 26, https://doi.org/10.1007/978-3-319-73126-1_2

Although this result can be extended to a ball in $\mathbb{C}^n$ (equipped with various specified norms) as well as to the Hilbert ball, this is no longer true in general Banach spaces. In the space $c_0 = \{x = (x_1, \ldots, x_n, \ldots), \ |x_i| \to 0 \text{ as } i \to \infty\}$ ($\|x\| = \max\limits_{1 \leq i < \infty} |x_i|$), there is a well-known example given by the self-mapping F of the open unit ball $\mathbb{B}$ defined by $F(x) = (a, x_1, x_2, \ldots)$ with $|a| \in (0, 1)$.

Note also that in the one-dimensional case, where $F \in \mathrm{Hol}(\Delta)$ is neither the identity mapping nor an elliptic automorphism of Δ, then the successful approximation method defined by the iterates $F^n = F \circ F^{n-1}$, $n = 2, 3, \ldots$, locally uniformly converges to the fixed point of F in Δ, whenever it exists. That is, the limit

$$\lim_{n \to \infty} F^n(x) = \tau$$

exists and does not depend on $x \in \Delta$ and $F(\tau) = \tau$. This point τ is called the Denjoy-Wolff point of F.

A finite-dimensional analog of this result is given in [17] (see also, [7, 16] and [28] and references therein).

However, again this fact is no longer true in the infinite-dimensional case, even when $X = H$ is a complex Hilbert space (see, for example, [7, 16, 31]). In this connection, see also the recent paper [3]. Therefore, another approximation method is needed for this general situation.

Observe also that since the early twentieth century many problems related to integro-differential equations as well as nonlinear operator equations (like Nekrasov and Hammestein equations in nonlinear mechanics and other applications, see, for example, [14, 32, 33] and [15]) have required the study of the fixed point set of holomorphic mappings in general infinite dimensional Banach spaces.

The simplest sufficient condition for $F \in \mathrm{Hol}(\mathbb{B}, X)$ to have an interior fixed point in $\mathbb{B}$ consists of the invariance condition

$$F(\mathcal{D}) \subset \mathcal{D} \tag{1}$$

for some closed convex subset $\mathcal{D} \subset \mathbb{B}$ and

$$\sup_{x \in \mathcal{D}} \left\| F'(x) \right\| < 1. \tag{2}$$

In the study of nonlinear operators having certain physical meaning, in order to satisfy conditions (1) and (2), one sometimes has to verify that some physical parameters are small enough (see [32, 33] and [15]).

However, conditions (1) and, of course, (2) are too strong and constrain us to use them for solving other problems.

Certainly, for integral operators (which are mostly relatively compact) one can use invariance condition (1) and the Leray-Schauder Theory under the additional restriction that F has a continuous extension to the boundary of $\mathcal{D}$ to ensure the existence but not always the uniqueness of the solution; see [14].

At the same time, condition (1) itself implies that $F \in \text{Hol}(\mathcal{D})$ is nonexpansive with respect to any metric on $\mathcal{D}$ given by the Schwarz-Pick system of pseudo-metrics on $\mathcal{D}$ [9]. (Note, in passing, that if $\mathcal{D} = \mathbb{B}$ then all such pseudo-metrics are the same and are actually metrics on $\mathbb{B}$; see [4].)

A breakthrough in this direction was made by Earle and Hamilton [5] (see also [7, 10, 11] and [28]) who proved that

Each holomorphic mapping of a bounded domain $\mathcal{D}$ in X into a subset $\Omega \subset \mathcal{D}$ that lies strictly inside D (i.e., Ω is bounded away from $\partial \mathcal{D}$) has a unique fixed point in $\mathcal{D}$. Moreover, if ρ is the Carathéodory-Riffen metric on $\mathcal{D}$, then F is a strict contraction with respect to this metric, i.e.,

$$\rho(F(x), F(y)) \leq q\rho(x, y), \quad x, y \in \mathcal{D}, \ 0 \leq q < 1.$$

However (even in the one-dimensional case) the assumptions of the Earle-Hamilton Theorem become not applicable if F does not map the open unit ball $\mathbb{B}$ into itself. Moreover, many examples show that F might be unbounded on $\mathbb{B}$ but still have a unique fixed point inside (see Example 6 below).

The goal of this paper is to assign such conditions that provide the existence and uniqueness of the fixed point in the open unit ball $\mathbb{B}$ of a general Banach space X for a mapping $F \in \text{Hol}(\mathbb{B}, X)$, which is not even necessarily bounded on $\mathbb{B}$.

Following Browder [2], Kirk and Schöneberg [13], and Morales [18, 19], we introduce the following notion.

Definition 1 A mapping $F \in \text{Hol}(\mathbb{B}, X)$ is said to be pseudo-contractive on $\mathbb{B}$ if for each $t \in [0, 1)$ and $y \in \mathbb{B}$, the equation

$$x = tF(x) + (1 - t)y \tag{3}$$

has a unique solution $x = \Phi_t(y) \in \mathbb{B}$ which is holomorphic in $y \in \mathbb{B}$.

In this connection, the papers by Reich [21, 22] are also relevant.

Note that if ρ is the hyperbolic (Kobayashi) metric on $\mathbb{B}$, then this solution $\Phi_t : \mathbb{B} \to \mathbb{B}$ is nonexpansive with respect to ρ:

$$\rho(\Phi_t(y_1), \Phi_t(y_2)) \leq q\rho(y_1, y_2) \tag{4}$$

for some $q \in [0, 1]$ and all y_1, y_2 in $\mathbb{B}$.

Definition 2 A mapping $F \in \text{Hol}(\mathbb{B}, X)$ is said to be strictly pseudo-contractive on $\mathbb{B}$ if condition (4) holds for all $y_1, y_2 \in \mathbb{B}$ with some $q \in [0, 1)$.

The point here is that actually condition (4) (with $0 \leq q \leq 1$) and the uniqueness of the solution of equation (3) imply that the fixed point sets of F and Φ_t in $\mathbb{B}$ are the same. Moreover, if F is strictly pseudo-contractive, i.e., $0 \leq q < 1$, then Φ_t (hence F) has a unique fixed point by the Banach Fixed Point Principle. In more general settings one can obtain the following assertion.

Lemma 3 *Let $F \in \mathrm{Hol}(\mathbb{B}, X)$ be a pseudo-contractive mapping on $\mathbb{B}$, and let $x_0 \in \mathbb{B}$ be a fixed point of F. If*

$$\mathrm{Ker}\left(I - F'(x_0)\right) \oplus \mathrm{Im}\left(I - F'(x_0)\right) = X,$$

then the fixed point set of F in $\mathbb{B}$ is a holomorphic retract of $\mathbb{B}$. Hence it is a connected submanifold of $\mathbb{B}$ tangent to $\mathrm{Ker}\left(I - F'(x_0)\right)$. In particular, if $I - F'(x_0)$ is an invertible linear operator, then x_0 is a unique fixed point of F in $\mathbb{B}$.

Proof It is sufficient to observe that for each $t \in [0, 1)$, the following equalities hold:

$$\mathrm{Ker}\left(I - F'(x_0)\right) = \mathrm{Ker}\left(I - (\Phi_t)'(x_0)\right) \quad \text{and}$$

$$\mathrm{Im}\left(I - F'(x_0)\right) = \mathrm{Im}\left(I - (\Phi_t)'(x_0)\right).$$

Then our assertion follows directly from Theorems 5.19 and 5.22 in [28]. □

It follows from the Earle-Hamilton Theorem that each holomorphic self-mapping of $\mathbb{B}$ is pseudo-contractive. Indeed, for each fixed $t \in [0, 1)$ and $y \in \mathbb{B}$, the mapping $G \in \mathrm{Hol}(\mathbb{B})$ defined by $G(x) = tF(x) + (1 - t)y$ maps $\mathbb{B}$ strictly inside itself, and this leads us to the conclusion.

However, even in the one-dimensional case, the class of pseudo-contractive mappings is much wider than the class of holomorphic self-mappings (see again Example 6 below). It even contains mappings which are unbounded in $\mathbb{B}$.

To formulate our main result, we need the following notions and notations.

As usual, the space of all continuous linear functionals on X is called the dual space of X and is denoted by X^*. We use the notation

$$\langle x, y^* \rangle := y^*(x)$$

for the semiscalar product on $X \times X^*$. We also denote by J the normalized duality mapping on X, i.e.,

$$J(x) := \left\{ x^* \in X^* : \langle x, x^* \rangle = \|x\|^2 = \|x^*\|^2 \right\}.$$

This, of course, is consistent with the case of the Hilbert space H, where H^* can be identified with H, and $\langle x, y \rangle$ with $y = y^*$ is the usual inner product in $H \times H$.

Theorem 4 *Let $\mathcal{D} = \mathbb{B}$ be the open unit ball in X, and let $F \in \mathrm{Hol}(\mathbb{B}, X)$. Assume that*

$$a = \sup_{\substack{x \in \partial \mathbb{B} \\ x^* \in J(x)}} \mathrm{Re}\left\langle F'(0)x, x^* \right\rangle < 1 \tag{5}$$

and for each $x \in \mathbb{B}$ and $x^* \in J(x)$

$$\mathrm{Re}\left\{F'(x)x, x^*\right\} + 2\|x\|^2 \|F(0)\| \le (1 + k)\|x\|^2, \qquad (6)$$

where

$$k \le \varkappa := (1 - a)\frac{2\log 2 - 1}{2(1 - \log 2)}. \qquad (7)$$

Then F is pseudo-contractive. Moreover, if the inequality in (7) is strong, then F is strictly pseudo-contractive, hence it has a unique fixed point.

Remark 5 If $F(0) = 0$, then the origin is a unique fixed point even when

$$k = \varkappa = (1 - a)\frac{2\log 2 - 1}{2(1 - \log 2)}.$$

Indeed, since $a < 1$ it follows that the linear operator $A = I - F'(0)$ is invertible.

Therefore, the origin is an isolated fixed point of F. On the other hand, the fixed point set of F in $\mathcal{D}$ is a holomorphic retract; hence it is a connected analytic submanifold of $\mathbb{B}$. Thus, it must contain the only fixed point.

In the one-dimensional case, where $X = \mathbb{C}$ and $\mathbb{B} = \Delta$ is the open unit disk in $\mathbb{C}$, the parameter a is just $\mathrm{Re}\, F'(0)$ and condition (6) can be rewritten as

$$\mathrm{Re}\, F'(z) < 1 + k - 2|F(0)|, \qquad (8)$$

where k satisfies (7).

Example 6 Consider the function $F \in \mathrm{Hol}(\Delta, \mathbb{C})$ defined as follows:

$$F(z) = 3(z - \log(1 + z)).$$

This function (as well as its real part) is not bounded in Δ. At the same time, $F(0) = 0$ and

$$F'(z) = 3\left(1 - \frac{1}{1 + z}\right) = \frac{3}{2}\left(1 - \frac{1 - z}{1 + z}\right).$$

Therefore, we have $a = \mathrm{Re}\, F'(0) = 0$ and

$$\mathrm{Re}\, F'(z) \le \frac{3}{2} < \frac{2\log 2 - 1}{2(1 - \log 2)}.$$

Thus, Theorem 4 implies that F is strictly pseudo-contractive; hence it has a pseudo-attractive fixed point $\tau \in \Delta$ in the sense that for each $t \in (0, 1)$, the limit

$$\lim_{n \to \infty} \Phi^n(t, z) = \tau,$$

where $\Phi(= \Phi_t(z))$ is the solution of equation (3).

Remark 7 We will see below (see Corollary 16) that actually the condition $\operatorname{Re} \langle F'(0)x, x^* \rangle \leq 1$ (cf. (5)), which means that the numerical range of the operator $A = F'(0)$ lies in the half plane $\Pi = \{z \in \mathbb{C} : \operatorname{Re} z \leq 1\}$, is necessary for $F \in \operatorname{Hol}(\mathbb{B}, X)$ to be pseudo-contractive.

To prove Theorem 4, we need the following assertion.

Lemma 8 *Let $\beta \in \mathbb{R}$ and Δ be the open unit disk in $\mathbb{C}$. If $p \in \operatorname{Hol}(\Delta, \mathbb{C})$ satisfies*

$$\operatorname{Re}(p(z) + zp'(z)) \geq \beta, \tag{9}$$

then

$$\operatorname{Re} p(z) \geq \beta + (\operatorname{Re} p(0) - \beta)\left(\frac{2\log(1 + |z|)}{|z|} - 1\right) \geq \varepsilon(\beta), \tag{10}$$

where $\varepsilon(\beta) := 2\beta(1 - \log 2) + \operatorname{Re} p(0) \cdot (2\log 2 - 1)$. The estimate is sharp.

Consequently, if $\operatorname{Re}(p(z) + zp'(z)) \geq \beta > \frac{1 - 2\log 2}{2(1 - \log 2)} \operatorname{Re} p(0)$, then $\operatorname{Re} p(z) \geq \varepsilon(\beta) > 0$.

Proof If inequality (9) holds, then the function q defined by

$$q(z) = \frac{p(z) + zp'(z) - i\operatorname{Im} p(0) - \beta}{\operatorname{Re} p(0) - \beta}$$

is of the Carathéodory class. Then

$$p(z) + zp'(z) = \beta + i\operatorname{Im} p(0) + (\operatorname{Re} p(0) - \beta)q(z) =: q_1(z). \tag{11}$$

Hence,

$$p(z) = \int_0^1 q_1(tz)\, dt = \beta + i\operatorname{Im} p(0) + (\operatorname{Re} p(0) - \beta)\int_0^1 q(tz)\, dt. \tag{12}$$

By Harnack's inequality,

$$\operatorname{Re} p(z) \geq \beta + (\operatorname{Re} p(0) - \beta) \int_0^1 \frac{1 - t|z|}{1 + t|z|} dt$$

$$= \beta + (\operatorname{Re} p(0) - \beta) \left(\frac{2 \log(1 + |z|)}{|z|} - 1 \right) \geq \varepsilon(\beta).$$

Therefore, estimate (10) holds. Choosing in (12) the function $q(z) = \dfrac{1 - z}{1 + z}$ and, consequently,

$$p(z) = \beta + i\gamma + (\operatorname{Re} p(0) - \beta) \int_0^1 \frac{1 - tz}{1 + tz} dt, \quad \gamma \in \mathbb{R},$$

we see that estimate (10) is sharp. $\qquad\qquad\square$

Now we are ready to prove Theorem 4.

Proof Fix $u \in \partial \mathbb{B}$ and $u^* \in J(u) \subset X^*$ and consider the function $g \in \operatorname{Hol}(\Delta, \mathbb{C})$ defined by

$$g(z)(= g_u(z)) = z - \langle F(zu), u^* \rangle + \langle F(0), u^* \rangle - \overline{\langle F(0), u^* \rangle} z^2. \tag{13}$$

Since $g(0) = 0$ one can write

$$g(z) = zp(z) \tag{14}$$

with $p \in \operatorname{Hol}(\Delta, \mathbb{C})$ and

$$\operatorname{Re} p(0) = \operatorname{Re} g'(0) = 1 - \operatorname{Re} \langle F'(0)u, u^* \rangle \geq 1 - a > 0.$$

Also, setting $x = zu$, $\|x\| = |z| \neq 0$, we have

$$\operatorname{Re} \left(p(z) + zp'(z) \right) = \operatorname{Re} g'(z)$$

$$= 1 - \operatorname{Re} \left(\langle F'(zu)u, u^* \rangle - 2z \overline{\langle F(0), u^* \rangle} \right)$$

$$\geq 1 - \operatorname{Re} \langle F'(x)x, x^* \rangle \frac{1}{\|x\|^2} - 2\|F(0)\|$$

$$= 1 - \frac{1}{\|x\|^2} \left(\operatorname{Re} \langle F'(x)x, x^* \rangle + 2\|x\|^2 \|F(0)\| \right)$$

$$\geq 1 - (1 + k) = -k \geq (1 - a) \frac{1 - 2 \log 2}{2(1 - \log 2)}.$$

Setting now $\beta = -k$ in Lemma 8, we have

$$\operatorname{Re} p(z) \geq \varepsilon(\beta) = -2k(1 - \log 2) + \operatorname{Re} p(0)(2 \log 2 - 1) \geq 0 \qquad (15)$$

whenever $-\frac{\operatorname{Re} p(0) \cdot (2 \log 2 - 1)}{2(1 - \log 2)} \leq \beta \leq \operatorname{Re} p(0)$, or, which is the same, $-\operatorname{Re} p(0) \leq k \leq \varkappa$.

In turn, it follows from (14) and (15) that for all $z \in \Delta$,

$$\operatorname{Re} g(z)\bar{z} = |z|^2 \operatorname{Re} p(z) \geq |z|^2 \varepsilon(-k) \geq 0.$$

On the other hand, by (13), we realize that

$$\operatorname{Re} g(z)\bar{z} = |z|^2 - \operatorname{Re} \langle F(zu), (zu)^* \rangle + \operatorname{Re} \langle F(0), u^* \rangle \bar{z} - \operatorname{Re} \overline{\langle F(0), u^* \rangle} z|z|^2$$
$$= \|x\|^2 - \operatorname{Re} \langle F(x), x^* \rangle + \operatorname{Re} \langle F(0), x^* \rangle \cdot (1 - \|x\|^2).$$

This implies that

$$\operatorname{Re} \langle F(x), x^* \rangle \leq \|x\|^2 + (1 - \|x\|^2)\| F(0) \| \|x\| - \varepsilon(-k)\|x\|^2$$
$$= \|x\|^2(1 - \varepsilon(-k)) + (1 - \|x\|^2)\| F(0) \| \|x\|. \qquad (16)$$

Now fix any $y \in \mathbb{B}$ and $t \in [0, 1)$ and consider the equation

$$x = tF(x) + (1 - t)y. \qquad (17)$$

Define the mapping $\Psi \in \operatorname{Hol}(\mathbb{B} \times \mathbb{B}, X)$ by

$$\Psi(x, y) = x - (tF(x) + (1 - t)y)$$

and set $\|x\| = r \in [0, 1)$. We have by (16) that

$$\operatorname{Re} \langle \Psi(x, y), x^* \rangle \geq r^2 - \left(t \operatorname{Re} \langle F(x), x^* \rangle + (1 - t)r\| y \| \right)$$
$$\geq r^2 - (tr^2 v + (1 - r^2)\| F(0) \|r + (1 - t)r\| y \|)$$
$$=: r\varphi(r),$$

where $v := 1 - \varepsilon(-k) < 1$ and

$$\varphi(r) = r - trv - (1 - r^2)\| F(0) \| - (1 - t)\| y \|$$
$$= r(1 - tv) - (1 - r^2)\| F(0) \| - (1 - t)\| y \|.$$

Since

$$\varphi(1) = 1 - tv - (1 - t)\| y \| = (1 - t)(1 - \| y \|) + t\varepsilon(-k) > 0, \qquad (18)$$

we obtain that for every fixed $y \in \mathbb{B}$ and $t \in [0, 1)$ there is $r_0 \in [0, 1)$ and $\delta > 0$ such that

$$\inf_{\|x\|=r_0} \text{Re}\, \langle \Psi(x, y), x^* \rangle > \delta. \tag{19}$$

Now it follows from a version of the implicit function theorem given in [23] (see also [28]) that equation $\Psi(x, y) = 0$, hence (17), has a unique solution $x = \Phi_t(y)$ with $\|x\| < r_0(= r_0(y, t)) < 1$, which holomorphically depends on $y \in \mathbb{B}$. Thus F is pseudo-contractive on $\mathbb{B}$.

Finally, if $k < \varkappa$, hence $\varepsilon(-k) > 0$, we have by (18) that inequality (19) holds for all $y \in \mathbb{B}$. Hence r_0 can be chosen independently of y and $t \in (0, 1]$. Since $\|\Phi_t(y)\| < r_0 < 1$, $y \in \mathbb{B}$, it follows from the Earle-Hamilton Theorem [5] that

$$\rho(\Phi_t(y_1), \Phi_t(y_2)) \leq q\rho(y_1, y_2), \quad y_1, y_2 \in \mathbb{B}, \tag{20}$$

with some $0 \leq q < 1$, i.e., F is strictly pseudo-contractive. The theorem is proved.
□

As we have already mentioned, if $q = 1$, that is, F is pseudo-contractive, but not strictly pseudo-contractive, the existence of a fixed point does not necessarily follow even if F is a self-mapping of $\mathbb{B}$ having a continuous extension to $\overline{\mathbb{B}}$. In addition, even if the fixed point set of F in $\overline{\mathbb{B}}$ is not empty, it is not clear whether the net $\{\Phi_t(y)\}_{t \in [0,1)}$ is pointwise (strongly) convergent to a fixed point of F when t tends to 1^- (compare with Eq. (3)). For a self-mapping of the Hilbert ball $\mathbb{B}$, the answer is affirmative (see [7] and references therein). We show that this fact holds in general as well. To formulate this theorem we need the following property of the solution $\Phi_t \in \text{Hol}(\mathbb{B})$.

Set $t = \frac{1}{2}$ and denote for simplicity $\Phi := \Phi_{\frac{1}{2}}$, i.e., $\Phi \in \text{Hol}(\mathbb{B})$ satisfies the equation

$$\Phi = \frac{1}{2}(F \circ \Phi + I). \tag{21}$$

Since Φ is a self-mapping of $\mathbb{B}$, then for each $s \in [0, 1)$ equation

$$G_s = s\Phi \circ G_s + (1 - s)I \tag{22}$$

has a solution $G_s \in \text{Hol}(\mathbb{B})$.

Lemma 9 *For all $t \in \left[\frac{1}{2}, 1\right)$, the following identity holds*

$$\Phi_t = \Phi \circ G_s,$$

where $s = 2 - \frac{1}{t} \in [0, 1)$.

Proof Since for $s = 0$ the assertion is trivial, we proceed with $s \in (0, 1)$. Let us rewrite Eqs. (21) and (22) in the form

$$I - \Phi = (I - F) \circ \Phi \tag{23}$$

and

$$\left[I + \frac{s}{1-s}(I - \Phi) \right] \circ G_s = I. \tag{24}$$

Substituting (23) into (24) we get the equality

$$G_s + \frac{s}{1-s}(I - F) \circ \Phi \circ G_s = I.$$

Denote $\Phi \circ G_s = \Psi_s \in \mathrm{Hol}(\mathbb{B})$ and write the last equality as

$$G_s + \frac{s}{1-s}(I - F) \circ \Psi_s = I. \tag{25}$$

On the other hand, by (22), we have $G_s = s\Psi_s + (1-s)I$. Comparing this relation with (25), we get

$$s\Psi_s + \frac{s}{1-s}(I - F) \circ \Psi_s = sI,$$

or, equivalently,

$$(1 - s)\Psi_s + (I - F) \circ \Psi_s = (1 - s)I,$$

or

$$\Psi_s = \frac{1}{2-s} F \circ \Psi_s + \frac{1-s}{2-s} I,$$

or

$$\Psi_s = tF \circ \Psi_s + (1 - t)I, \tag{26}$$

where $t = \frac{1}{2-s} \in \left[\frac{1}{2}, 1 \right)$ wherever $s \in [0, 1)$.

Finally, since (26) has a unique solution that by definition is Φ_t we have $\Phi_t = \Psi_s := \Phi \circ G_s$ and we are done. $\square$

Theorem 10 *Let* $\mathbb{B}$ *be the open unit ball in a complex Hilbert space H. Suppose that $F \in \text{Hol}(\mathbb{B}, H)$ is pseudo-contractive, that is, the equation*

$$\Phi_t = tF \circ \Phi_t + (1-t)I, \quad t \in [0,1),$$

has a holomorphic solution $\Phi_t \in \text{Hol}(\mathbb{B})$ for each $t \in [0,1)$. Then for each $x \in \mathbb{B}$ the curve $\Phi_t(x)$ converges to a point $\widetilde{x} \in \overline{\mathbb{B}}$.

Proof Denote by $\{G_s\}_{s \in [0,1)}$ the net defined by (22) and fix $x \in \mathbb{B}$. Then $\{G_s(x)\}$ is an approximative curve in $\mathbb{B}$. This curve converges to a point $a \in \overline{\mathbb{B}}$ as $s \to 1^-$ (see [7]). Also we have by (22),

$$\|G_s(x) - (\Phi \circ G_s)(x)\| = \|(1-s)(I - \Phi \circ G_s)(x)\|$$

$$\leq (1-s)\left[\|\Phi \circ G_s(x)\| + \|x\|\right] \leq 2(1-s),$$

which tends to zero as $s \to 1^-$. By Lemma 9, this fact completes the proof of Theorem 10. □

Now we consider the problems concerning the common fixed point set of a family of mappings that forms a one-parameter continuous semigroup.

Definition 11 A family $S = \{F_t\}_{t \geq 0} \subset \text{Hol}(\mathcal{D})$ is called a one-parameter continuous semigroup on $\mathcal{D}$ if the following properties hold:

(i) $F_{t+s} = F_t \circ F_s$ for all $t, s \geq 0$;
(ii) $\lim_{t \to 0^+} F_t(x) = x$, where the limit is taken with respect to the strong topology of X.

If the last limit is uniform on each subset strictly inside $\mathcal{D}$ (bounded away from the boundary of $\mathcal{D}$), then one says that the semigroup is locally uniformly continuous in $\mathcal{D}$.

It is known that each one-parameter locally uniformly continuous semigroup on a bounded convex domain is differentiable with respect to the parameter $t \geq 0$; see [23]. Therefore, the limit

$$f = \lim_{t \to 0^+} \frac{1}{t}[F_t - I] \tag{27}$$

exists in the topology of locally uniform convergence on $\mathcal{D}$. The mapping f is called the (infinitesimal) generator of the semigroup S. In this case the semigroup S can be reproduced as the solution of the Cauchy problem

$$\begin{cases} \frac{\partial u(t,x)}{\partial t} + f(u(t,x)) = 0 \\ u(0,x) = x, \end{cases} \tag{28}$$

where we set $F_t(x) = u(t,x)$.

It follows from the uniqueness of the solution to the Cauchy problem (28) that the common fixed point set of S coincides with the null point set of its generator f.

Definition 12 A mapping $f \in \mathrm{Hol}(\mathcal{D}, X)$ is called a semi-complete vector field if for every $x \in \mathcal{D}$, the Cauchy problem (28) has a unique solution $u = u(t, x) \in \mathcal{D}$ for all $t \geq 0$.

Thus, for a bounded convex domain, the notions of semi-complete vector field and infinitesimal generator are the same. Moreover, one can establish the following assertion, which we formulate for the case where $\mathcal{D} = \mathbb{B}$ is the open unit ball in X.

Proposition 13 Let $f \in \mathrm{Hol}(\mathbb{B}, X)$. The following assertions are equivalent:

(i) f is a semi-complete vector field;

(ii) there is a positive $T > 0$ and a family $\{\mathcal{F}_t\}_{t\in(0,T)} \subset \mathrm{Hol}(\mathbb{B})$ such that

$$\lim_{t\to 0^+} \frac{x - \mathcal{F}_t(x)}{t} = f(x)$$

for each $x \in \mathbb{B}$;

(iii) the mapping f satisfies the range condition

$$(I + rf)(\mathbb{B}) \supseteq \mathbb{B}, \ r \geq 0,$$

and for each $r \geq 0$

$$(I + rf)^{-1} \in \mathrm{Hol}(\mathbb{B});$$

(iv) the mapping f satisfies the condition

$$\mathrm{Re}\,\langle f(x), x^* \rangle \geq \mathrm{Re}\,\langle f(0), x^* \rangle\,(1 - \|x\|^2), \quad x \in \mathbb{B}, \ x^* \in J(x);$$

(v) the mapping $F = I - f$ is pseudo-contractive.

The mapping $(I + rf)^{-1}$ in assertion (iii) is called the nonlinear resolvent for f.

Proof Since the equivalence of the first three assertions was proved in [26] and the equivalence (i)$\Longleftrightarrow$(iv) was proved in [28] (see also [1, 12, 24]), it suffices to prove that (i) is equivalent to (v).

If F is pseudo-contractive, then we set $\mathcal{F}_t = \Phi_t$ where Φ_t is the solution of the equation

$$\Phi_t(x) = tF(\Phi_t(x)) + (1 - t)x. \tag{29}$$

We have

$$\frac{x - \mathcal{F}_t(x)}{t} = \frac{1}{t}[x - tF(\mathcal{F}_t(x)) - (1 - t)x] = x - F(\mathcal{F}_t(x)). \tag{30}$$

Since $\mathcal{F}_t(x)$ continuously depends on t, we get that there is $y \in \mathbb{B}$ such that

$$\lim_{t \to 0^+} \mathcal{F}_t(x) = y$$

and $y = x$ by (29). Thus by (30) we obtain that

$$\lim_{t \to 0^+} \frac{x - \mathcal{F}_t(x)}{t} = f(x) := x - F(x).$$

Conversely. Let $f = I - F$ be a semi-complete vector field. Then by assertion (iii) we conclude that the equation

$$x + r(x - F(x)) = y \tag{31}$$

has a unique solution $x = x(r, y)$ for each pair $r \geq 0$ and $y \in \mathbb{B}$, holomorphically depending on $y \in \mathbb{B}$. Setting here $r = \frac{t}{1-t}$, $t \in [0, 1)$, one transforms Eq. (31) to the form

$$x = tF(x) + (1 - t)y,$$

and we are done. □

As we have mentioned above, the common fixed point set of a locally uniformly continuous semigroup coincides with the null point set of its infinitesimal generator. Therefore, Theorem 4 and Proposition 13 immediately imply the following result.

Corollary 14 *Let $f \in \mathrm{Hol}(\mathbb{B}, X)$ satisfy the conditions*

$$a = \inf_{\substack{x \in \partial \mathbb{B} \\ x^* \in J(x)}} \mathrm{Re}\langle f'(0)x, x^* \rangle > 0$$

and

$$\mathrm{Re}\langle f'(x)x, x^* \rangle - 2\|x\|^2 \|f(0)\| \geq -k\|x\|^2$$

for some $k \leq \varkappa := (1 - a)\dfrac{2\log 2 - 1}{2(1 - \log 2)}$. Then f is a semi-complete vector field. Moreover, if $k < \varkappa$ then the semigroup $S = \{F_t\}_{t \geq 0}$ generated by f has a unique common fixed point x in $\mathbb{B}$. In addition, if $f(0) = 0$, then the following global uniform estimate holds:

$$\| F_t(x) \| \leq e^{-bt} \|x\|,$$

where $b = -2(1 - \log 2)k + (2\log 2 - 1)(1 - a) > 0$.

In its turn, Corollary 14 enables us to prove a new rigidity theorem. To explain it, we observe the following clear fact: if a real function F is continuous in $[0, 1]$, preserves the endpoints of this segment, is smooth inside it and satisfies $F'(x) \leq 1$, then it must be the identity mapping. We show that for functions that are analytic in the open unit disk in $\mathbb{C}$, the last inequality can be weakened in a sense.

Theorem 15 *Let $F \in \mathrm{Hol}(\Delta, \mathbb{C})$ satisfy $F(0) = 0$, $\mathrm{Re}\, F'(0) \leq 1$ and*

$$\mathrm{Re}\, F'(z) \leq 1 + \frac{2 \log 2 - 1}{2(1 - \log 2)}(1 - \mathrm{Re}\, F'(0)), \quad z \in \Delta.$$

Assume that either one of the following conditions holds:

(i) *the radial derivative $F'(1) = \lim\limits_{r \to 1^-} \dfrac{F(r) - 1}{r - 1}$ exists with $\mathrm{Re}\, F'(1) \leq 1$;*
or

(ii) $F'(0) = 1$.

Then $F(z) \equiv z$.

Proof Define a sequence of functions $\{f_n\}$ by $f_n(z) = \frac{n+1}{n} z - F(z)$. These functions satisfy the conditions $f_n = 0$, $\mathrm{Re}\, f_n'(0) > 0$ and

$$\mathrm{Re}\, f_n'(z) \geq \frac{1}{n} \left(\frac{2 \log 2 - 1}{2(1 - \log 2)} + 1 \right) - \frac{2 \log 2 - 1}{2(1 - \log 2)} \, \mathrm{Re}\, f_n'(0)$$

$$> -\frac{2 \log 2 - 1}{2(1 - \log 2)} \, \mathrm{Re}\, f_n'(0).$$

Hence, by Corollary 14, each function f_n is a semi-complete vector field. Since the set of all semi-complete vector fields in the disk Δ is a closed cone (see, for example, [1, 30]), the limit of this sequence, that is, the function $f(z) = \lim\limits_{n \to \infty} f_n(z) = z - F(z)$ is a semi-complete vector field too.

If condition (i) holds, then $f(1) = 0$ and $\mathrm{Re}\, f'(1) \geq 0$. Therefore, if f is not identically zero, then the boundary point 1 is the attractive fixed point for the semigroup generated by f. In this case, f cannot have interior null points. This contradicts the fact that $f(0) = 0$. Therefore, $f(z) = z - F(z) \equiv 0$.

If condition (ii) holds, then $f(0) = f'(0) = 0$, that is, f vanishes at the point $z_0 = 0$ at least up to the second order. This may happen only when once again $f \equiv 0$. The proof is complete. $\square$

In addition, Proposition 13 can be used to obtain more detailed information on pseudo-contractive mappings.

Theorem 16 *Let F be a pseudo-contractive mapping on $\mathbb{B}$. Then the numerical range of the linear operator $A = F'(0)$ lies in the half-plane $\Pi = \{z \in \mathbb{C} : \mathrm{Re}\, z \leq 1\}$. In other words, for each $x \in \partial \mathbb{B}$ and $x^* \in J(x)$,*

$$\mathrm{Re}\, \langle Ax, x^* \rangle \leq 1.$$

Proof Consider the mapping $f = I - F$, which by the equivalence (i)$\Longleftrightarrow$(v) of Proposition 13 is a semi-complete vector field on $\mathbb{B}$, and hence by (iv) satisfies the condition

$$\text{Re}\,\langle f(x), x^* \rangle \geq \text{Re}\,\langle f(0), x^* \rangle\,(1 - \|x\|^2), \quad x \in \mathbb{B},\ x^* \in J(x). \tag{32}$$

For any $u \in \partial\mathbb{B}$ and $u^* \in J(u)$, consider the function $f_u \in \text{Hol}(\Delta, \mathbb{C})$ defined by $f_u(z) = \langle f(zu), u^* \rangle$. Inequality (32) implies that f_u can be written in the form

$$f_u(z) = f_u(0) - \overline{f_u(0)}z^2 + z \cdot q(z)$$

with $q \in \text{Hol}(\Delta, \mathbb{C})$, $\text{Re}\,q(z) \geq 0$.
In its turn, this implies that

$$(f_u)'(0) = \lim_{z \to 0} \frac{f_u(z) - f_u(0)}{z} = -\lim_{z \to 0}\left[\overline{f_u(0)}z + q(z)\right] = q(0).$$

Now returning to our notations and settings $x = zu$, $x^* = \bar{z}u^*$, we have

$$\text{Re}\,\langle f'(0)x, x^* \rangle = \text{Re}\,\langle f'(0)zu, (zu^*) \rangle = \text{Re}\,q(0)|z|^2 \geq 0,$$

which means that

$$\text{Re}\,\langle Ax, x^* \rangle = \text{Re}\,\langle (I - f'(0))x, x^* \rangle \leq 1,$$

and we are done. □

We complete the paper with the following result on extension operators for pseudo-contractive mappings in the spirit of Pfaltzgraff and Suffridge; see [20].

Theorem 17 *Let $X = \mathbb{C}^n$ be the n-dimensional Euclidean space and let $\mathbb{B}_n \subset \mathbb{C}^n$ be the open unit ball. Let Y be any complex Banach space and the space $\mathbb{C}^n \times Y$ be equipped with the norm*

$$\|(x, y)\| = \left(\|x\|^2 + \|y\|^2\right)^{1/2}.$$

If F is a pseudo-contractive mapping on $\mathbb{B}_n$, then the mapping G, which is holomorphic in the open unit ball $\mathbb{B}$ of the space $\mathbb{C}^n \times Y$ and defined by

$$G(x, y) = \left(F(x), \frac{1 + \text{trace}\, F'(x)}{n + 1}\, y \right),$$

is pseudo-contractive on $\mathbb{B}$. Moreover, if F is strictly pseudo-contractive, then G has a unique fixed point in $\mathbb{B}$.

Proof Since F is pseudo-contractive, the mapping $f = I - F$ is a semi-complete vector field by Proposition 13. Denote by $\{F_t\}_{t \geq 0}$ the semigroup generated by f. Then by Elin [6, Theorem 4.1 and Example 1], the family $\{G_t\}$ defined by

$$G_t(x, y) = \left(F_t(x), \left(\det F_t'(x) \right)^{\frac{1}{n+1}} y \right)$$

forms a one-parameter continuous semigroup on the ball $\mathbb{B} \subset \mathbb{C}^n \times Y$. To find its generator $g \in \text{Hol}(\mathbb{B}, \mathbb{C}^n \times Y)$, we just differentiate $G_t(x, y)$ at $t = 0$ using regular rules of differentiation of determinants:

$$\frac{\partial}{\partial t} \det F_t'(x) \Big|_{t=0} = -\operatorname{trace} f'(x).$$

Hence,

$$g(x, y) = -\frac{\partial}{\partial t} G_t(x, y) \Big|_{t=0} = \left(f(x), \frac{\operatorname{trace} f'(x)}{n+1} y \right).$$

Using Proposition 13 again, we conclude that the mapping $G := I - g$ is pseudo-contractive on the unit ball $\mathbb{B}$.

If F is strictly pseudo-contractive, then it has a unique fixed point $x_0 \in \mathbb{B}$. Then there is $q \in (0, 1)$ such that

$$\operatorname{Re} \langle F'(x_0)x, x^* \rangle < q \qquad (33)$$

for all $x \in \partial\mathbb{B}$ and $x^* \in J(x)$. In particular, this implies that $\operatorname{Re} \frac{\partial F_j(x_0)}{\partial x_j} < q$ for every $j = 1, \ldots, n$.

Obviously, the point $(x_0, 0)$ is a fixed point of G. We have to show that it is the only fixed point of G. Consider the linear operator $A = G'(x_0, 0)$. A straightforward calculation shows that it acts as follows:

$$A(x, y) = \left(F'(x_0)x, \frac{1 + \operatorname{trace} F'(x_0)}{n+1} y \right),$$

and hence, for any point (x, y) on the unit sphere of the space $\mathbb{C}^n \times Y$ with $x \neq 0$, we have

$$\langle A(x, y), (x, y)^* \rangle = \left\langle F'(x_0) \frac{x}{\|x\|}, \frac{x^*}{\|x\|} \right\rangle \|x\|^2 + \frac{1 + \operatorname{trace} F'(x_0)}{n+1} (1 - \|x\|^2).$$

Therefore, (33) implies that

$$\operatorname{Re} \langle A(x, y), (x, y)^* \rangle < q\|x\|^2 + \frac{1 + nq}{n+1} (1 - \|x\|^2) \leq \frac{1 + nq}{n+1} < 1.$$

Thus, the fixed point $(x_0, 0)$ is isolated, and hence unique. The proof is complete.

$\square$

Remark 18 It was shown in [6] that the Pfaltzgraff–Suffridge extension operator can serve to extend one-parameter semigroups. Using a similar scheme, one can use various extension operators for semigroups and obtain other extension operators for pseudo-contractive mappings. For instance, using the operator $f(x) \rightarrow \left(f(x), \left(\frac{f(x)}{x} \right)^{\beta} y \right)$ with $\beta > 0$ introduced in [8], one shows that if F is pseudo-contractive on the open unit disk $\Delta \subset \mathbb{C}$ and $F(0) = 0$, then the mapping $G \in \text{Hol}(\mathbb{B}, \mathbb{C} \times Y)$ defined by

$$G(x, y) = \left(F(x), \left(1 - \beta + \beta \frac{F(x)}{x} \right) y \right)$$

is pseudo-contractive. Moreover, it has the unique fixed point, provided F is strictly pseudo-contractive.

Example 19 We have already seen in Example 6 that the function $F \in \text{Hol}(\Delta, \mathbb{C})$ defined by

$$F(z) = 3(z - \log(1 + z))$$

is a strictly pseudo-contractive mapping. Let Y be an arbitrary Banach space and β be a positive number. Then by Theorem 17 and Remark 18 the mappings G_1, G_2 holomorphic in the open unit ball $\mathbb{B} := \{(z, y) \in \mathbb{C} \times Y : |z|^2 + \|y\|^2 < 1\}$ and defined by

$$G_1(z, y) = \left(3(z - \log(1 + z)), \frac{1 + 4z}{2(1 + z)} y \right)$$

and

$$G_2(z, y) = \left(3(z - \log(1 + z)), \left(1 + 2\beta - 3\beta \frac{\log(1 + z)}{z} \right) y \right)$$

are pseudo-contractive on $\mathbb{B}$ and have a unique fixed point.

Acknowledgements The work was partially supported by the European Commission under the project STREVCOMS PIRSES-2013-612669. The publication was prepared with the support of the "RUDN University Program 5-100". Both authors are grateful to the anonymous referee for the very fruitful remarks.

References

1. Aharonov, D., Elin, M., Reich, S., Shoikhet, D.: Parametric representations of semi-complete vector fields on the unit balls in $\mathbb{C}^n$ and in Hilbert space. Atti Accad. Naz. Lincei **10**, 229–253 (1999)
2. Browder, F.E.: Nonlinear mappings of nonexpansive and accretive type in Banach spaces. Bull. Am. Math. Soc. **73**, 875–882 (1967)
3. Budzyńska, M., Kuczumow, T., Reich, S.: The Denjoy-Wolff iteration property in complex Banach spaces. J. Nonlinear Convex Anal. **17**, 1213–1221 (2016)
4. Dineen, S.: The Schwartz Lemma. Clarendon, Oxford (1989)
5. Earle, C.J., Hamilton, R.S.: A fixed point theorem for holomorphic mappings. In: Proceedings of Symposia in Pure Mathematics, vol. 16, pp. 61–65. American Mathematical Society, Providence, RI (1970)
6. Elin, M.: Extension operators via semigroups. J. Math. Anal. Appl. **377**, 239–250 (2011)
7. Goebel, K., Reich, S.: Uniform convexity, hyperbolic geometry, and nonexpansive mappings. In: Monographs and Textbooks in Pure and Applied Mathematics, vol. 83. Marcel Dekker, New York (1984)
8. Graham, I., Kohr, G.: Univalent mappings associated with the Roper-Suffridge extension operator. J. Anal. Math. **81**, 331–342 (2000)
9. Harris, L.A.: Schwarz-Pick systems of pseudometrics for domains in normed linear spaces. In: Advances in Holomorphy, pp. 345–406. North Holland, Amsterdam (1979)
10. Harris, L.A.: Fixed points of holomorphic mappings for domains in Banach spaces. Abstr. Appl. Anal. **5**, 261–274 (2003)
11. Harris, L.A.: Fixed point theorems for infinite dimensional holomorphic functions. J. Korean Math. Soc. **41**, 175–192 (2004)
12. Khatskevich, V., Reich, S., Shoikhet, D.: Complex dynamical systems on bounded symmetric domains. Electron. J. Differ. Equ. **19**, 1–9 (1997)
13. Kirk, W.A., Schöneberg, R.: Some results on pseudocontractive mappings. Pac. J. Math. **71**, 89–100 (1977)
14. Krasnosel'skiĭ, M.A., Zabreĭko, P.P.: Geometrical Methods of Nonlinear Analysis. Grundlehren der Mathematischen Wissenschaften, vol. 263. Springer, Berlin (1984)
15. Krasnosel'skiĭ, M.A., Vaĭnikko, G.M., Zabreĭko, P.P., Rutitskii, Ya.B., Stecenko, V.Ya.: Approximate Solution of Operator Equations. Wolters-Noordhoff, Groningen (1972)
16. Kuczumow, T., Reich, S., Shoikhet, D.: Fixed points of holomorphic mappings: a metric approach. In: Handbook of Metric Fixed Point Theory, pp. 437–515. Kluwer, Dordrecht (2001)
17. MacCluer, B.D.: Iterates of holomorphic self-maps of the unit ball in $\mathbb{C}^n$. Michigan Math. J. **30**, 97–106 (1983)
18. Morales, C.: Pseudo-contractive mappings and the Leray-Schauder boundary condition. Comment. Math. Univ. Carol. **4**, 745–756 (1979)
19. Morales, C.: Remarks on pseudo-contractive mappings. J. Math. Anal. Appl. **87**, 158–164 (1982)
20. Pfaltzgraff, J.A., Suffridge, T.J.: An extension theorem and linear invariant families generated by starlike maps. Ann. Univ. Mariae Curie-Sk lodowska Sect. A **53**, 193–207 (1999)
21. Reich, S.: Minimal displacement of points under weakly inward pseudo-lipschitzian mappings. Atti Accad. Naz. Lincei **59**, 40–44 (1975)
22. Reich, S.: On the fixed point theorems obtained from existence theorems for differential equations. J. Math. Anal. Appl. **54**, 26–36 (1976)
23. Reich, S., Shoikhet, D.: Generation theory for semigroups of holomorphic mappings in Banach spaces. Abstr. Appl. Anal. **1**, 1–44 (1996)
24. Reich, S., Shoikhet, D.: Semigroups and generators on convex domains with the hyperbolic metric. Atti Accad. Naz. Lincei **8**, 231–250 (1997)
25. Reich, S., Shoikhet, D.: The Denjoy-Wolff theorem. Ann. Univ. Mariae Curie-Sk lodowska Sect. A **51**, 219–240 (1997)

26. Reich, S., Shoikhet, D.: Averages of holomorphic mappings and holomorphic retractions on convex hyperbolic domains. Stud. Math. **130**, 231–244 (1998)
27. Reich, S., Shoikhet, D.: The Denjoy–Wolff theorem. Encyclopedia of Mathematics, Supplement 3, pp. 121–123. Kluwer, Dordrecht (2002)
28. Reich, S., Shoikhet, D.: Nonlinear Semigroups, Fixed Points, and the Geometry of Domains in Banach Spaces. Imperial College, London (2005)
29. Shapiro, J.H.: Composition Operators and Classical Function Theory. Springer, Berlin (1993)
30. Shoikhet, D.: Semigroups in Geometrical Function Theory. Kluwer, Dordrecht (2001)
31. Stachura, A.: Iterates of holomorphic self-maps of the unit ball in Hilbert space. Proc. Am. Math. Soc. **93**, 88–90 (1985)
32. Trenogin, V.A.: Functional Analysis. Nauka, Moscow (1980)
33. Vainberg, M.M., Trenogin, V.A.: Theory of Bifurcation of Solutions of Nonlinear Equations. Nauka, Moscow (1969)

On Parabolic Dichotomy

Leandro Arosio

Abstract If f is a parabolic holomorphic self-map of the unit disc $\mathbb{D} \subset \mathbb{C}$, then either for every point z one has $\lim_{n \to \infty} k_{\mathbb{D}}(f^{n+1}(z), f^n(z)) > 0$, or the Poincaré distance of any two f-orbits converges to zero. It is an open question whether such a dichotomy holds in the unit ball $\mathbb{B}^q \subset \mathbb{C}^q$. We show how this question is related to the theory of canonical Kobayashi hyperbolic semi-models, to commuting holomorphic self-maps of the ball and to a purely geometric problem about biholomorphisms of the ball.

Keywords Holomorphic dynamics · Models · Iteration

2010 Mathematics Subject Classification 32H50, 39B12, 26A18

1 Preliminaries

Holomorphic dynamics studies the behaviour of the iterates $(f^n)_{n \in \mathbb{N}}$ of a holomorphic self-map $f \colon X \to X$ of a complex manifold X. We are interested in the case $X = \mathbb{B}^q \subset \mathbb{C}^q$, the unit ball. In $\mathbb{B}^q$ the Kobayashi pseudodistance $k_{\mathbb{B}^q}$ (which is identically 0 in other interesting cases such as $\mathbb{P}^q(\mathbb{C})$ or $\mathbb{C}^q$) is a complete distance inducing the euclidean topology, for which any holomorphic self-map $f \colon \mathbb{B}^q \to \mathbb{B}^q$ is non-expansive. This prevents chaotic behaviour.

The following result by Hervé [14] is a generalization of the Denjoy–Wolff theorem in the unit disc, and shows that when it comes to dynamics in $\mathbb{B}^q$ also the points at the boundary $\partial \mathbb{B}^q$ play an important role.

L. Arosio (✉)
Dipartimento Di Matematica, Università di Roma " Tor Vergata", Roma, Italy
e-mail: arosio@mat.uniroma2.it

© Springer International Publishing AG, part of Springer Nature 2017   31
F. Bracci (ed.), *Geometric Function Theory in Higher Dimension*,
Springer INdAM Series 26, https://doi.org/10.1007/978-3-319-73126-1_3

Theorem 1.1 *Let $f: \mathbb{B}^q \to \mathbb{B}^q$ be a holomorphic map. Assume that f admits no fixed points in $\mathbb{B}^q$. Then there exists a point $p \in \partial\mathbb{B}^q$ such that $f^n \to p$ uniformly on compact subsets.*

The point p is called the Denjoy–Wolff point of f. The number

$$\lambda(f) := \liminf_{z \to p} \frac{1 - \|f(z)\|}{1 - \|z\|}$$

satisfies $0 < \lambda(f) \leq 1$ and is called the *dilation* of f (at p).

Thanks to this result one can divide the holomorphic self-maps of the ball in three natural families.

Definition 1.2 Let $f: \mathbb{B}^q \to \mathbb{B}^q$ be a holomorphic map.

(1) We call f elliptic if it admits a fixed point $z \in \mathbb{B}^q$.
(2) We call f parabolic if it is not elliptic and if its dilation $\lambda(f)$ is 1.
(3) We call f hyperbolic if it is not elliptic and if its dilation satisfies $0 < \lambda(f) < 1$.

Moreover, for all $z \in \mathbb{B}^q$, we define the *step* of f at z by

$$\text{step}_f(z) := \lim_{n \to \infty} k_{\mathbb{B}^q}(f^n(z), f^{n+1}(z)),$$

which is well-defined and finite by the non-expansiveness of the Kobayashi distance $k_{\mathbb{B}^q}$. The map f is said to be *zero-step* if $\text{step}_f(z) = 0$ for all $z \in \mathbb{B}^q$, and it is said to be *nonzero-step* if $\text{step}_f(z) > 0$ for all $z \in \mathbb{B}^q$.

Example 1.3 Let F be an automorphism of the unit disc $\mathbb{D}$. Let $\mathbb{H}$ denote the upper half-plane.

(1) If F is elliptic, then it is conjugate to the self-map $z \mapsto e^{i\vartheta}z$ of $\mathbb{D}$, where $\vartheta \in \mathbb{R}$.
(2) If F is parabolic, then it is conjugate to the self-map $z \mapsto z \pm 1$ of the upper-half-plane $\mathbb{H}$.
(3) If F is hyperbolic, then it is conjugate to the self-map $z \mapsto \frac{1}{\lambda(F)}z$ of the upper-half-plane $\mathbb{H}$.

Since the dynamics of the automorphisms of the ball is very simple to understand, it is natural to search for a semiconjugacy of the map $f: \mathbb{B}^q \to \mathbb{B}^q$ with an automorphism of a possibly lower-dimensional ball $\mathbb{B}^d$, or at least with an automorphism of some "simple" complex manifold. This motivates the following definition.

Definition 1.4 Let $f: \mathbb{B}^q \to \mathbb{B}^q$ be a holomorphic map. A *semi-model* for f is a triple (Λ, h, φ), where Λ is a complex manifold called the *base space*, φ is an automorphism of Λ and $h: \mathbb{B}^q \to \Lambda$ is a holomorphic map called the *intertwining map*, such that

$$h \circ f = \varphi \circ h,$$

and such that

$$\Lambda = \bigcup_{n \in \mathbb{N}} \varphi^{-n}(h(\mathbb{B}^q)).$$

We first consider the case of the unit disc. In the following theorem we recall classical results of Königs [16], Valiron [20], Pommerenke [18], and Baker–Pommerenke [6]. We follow the presentation of Cowen [10], who unified such results in a common framework.

Theorem 1.5 *Let $f \colon \mathbb{D} \to \mathbb{D}$ be a holomorphic self-map. Then there exists an f-invariant domain $U \subset \mathbb{D}$ such that for all $z \in \mathbb{D}$ the orbit $(f^n(z))$ eventually lies in U and such that $f|_U$ is injective. In the following cases there exists a semi-model (Ω, h, ψ), where $h \colon \mathbb{D} \to \mathbb{C}$ is injective on U.*

  (i) *[Königs] If f is an elliptic map with a fixed point p such that $0 < |f'(p)| < 1$, then $\Omega = \mathbb{C}$ and $\psi(z) = f'(p)z$.*
 (ii) *[Baker–Pommerenke] If f is a parabolic map with a point of zero step, then $\Omega = \mathbb{C}$ and $\psi(z) = z + 1$.*
(iii) *[Pommerenke] If f is a parabolic map with a point of non-zero step, then $\Omega = \mathbb{H}$ and $\psi(z) = z \pm 1$, which is a parabolic automorphism of $\mathbb{H}$.*
 (iv) *[Valiron] If f is a hyperbolic map, then $\Omega = \mathbb{H}$ and $\psi(z) = \frac{1}{\lambda(f)}z$, which is a hyperbolic automorphism of $\mathbb{H}$ with dilation $\lambda(f)$.*

Remark 1.6 If a parabolic holomorphic self-map $f \colon \mathbb{D} \to \mathbb{D}$ has zero step at a point $z \in \mathbb{D}$, then it has zero step at all points in $\mathbb{D}$. Actually we can say more: the Poincaré distance of every two orbits converges to zero. Indeed, for all $z, w \in \mathbb{D}$,

$$\lim_{n \to \infty} k_{\mathbb{D}}(f^n(z), f^n(w)) = k_{\mathbb{C}}(h(z), h(w)) = 0.$$

In what follows we refer to this property as *parabolic dichotomy*.

2 Canonical Semi-Models

We now consider the case of the unit ball $\mathbb{B}^q \subset \mathbb{C}^q$. Using the Poincaré–Dulac method (see e.g. [19]) one gets that if f is an elliptic map with a fixed point p such that the eigenvalues $\lambda_1, \ldots, \lambda_q$ of $d_p f$ satisfy $0 < |\lambda_i| < 1$ for all $1 \leq i \leq q$, then there exists a semi-model $(\mathbb{C}^q, h, \psi)$, where ψ is a triangular polynomial automorphism of $\mathbb{C}^q$, and h is a local biholomorphism near p.

Bracci–Gentili–Poggi-Corradini [9] studied the case of a hyperbolic holomorphic self-map $f \colon \mathbb{B}^q \to \mathbb{B}^q$, and, assuming some regularity at p, they proved the existence of a one-dimensional semi-model $(\mathbb{H}, h, z \mapsto \frac{1}{\lambda(f)}z)$ for f. For other results about semi-models in the unit ball, see [7, 8, 15].

Recently, Bracci and the author [2, 3] introduced an abstract approach to the construction of semi-models for self-maps of the ball $\mathbb{B}^q$ much in the spirit of the work of Cowen in the unit disc, obtaining the following result which generalizes parts (iii) and (iv) of Theorem 1.5.

Theorem 2.1 *Let* $f: \mathbb{B}^q \to \mathbb{B}^q$ *be holomorphic. Then there exists a semi-model* $(\mathbb{B}^d, \ell, \tau)$ *which satisfies the following universal property: if* (Λ, j, φ) *is a semi-model for* f *with Kobayashi hyperbolic base space, then there exists a unique holomorphic mapping* $\eta: \mathbb{B}^d \to \Lambda$ *such that the following diagram commutes:*

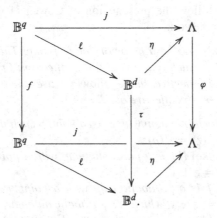

Moreover,

(1) *if* f *is hyperbolic, then* $d > 0$ *and* τ *is a hyperbolic automorphism with dilation* $\lambda(f)$,
(2) *if* f *is parabolic non-zero step, then* $d > 0$ *and* τ *is a parabolic automorphism.*

Remark 2.2 It follows immediately from the universal property that the semi-model $(\mathbb{B}^d, \ell, \tau)$ is essentially unique. We call it the *canonical Kobayashi hyperbolic semi-model* of f (canonical semi-model for brevity). The Kobayashi distance in the canonical semi-model is related to the dynamics of f in the following way

$$k_{\mathbb{B}^d}(\ell(z), \ell(w)) = \lim_{n \to \infty} k_{\mathbb{B}^q}(f^n(z), f^n(w)), \quad \forall z, w \in \mathbb{B}^q. \tag{1}$$

The canonical semi-model may have dimension 0, that is, it may reduce to a point. It follows from (1) this happens if and only if

$$k_{\mathbb{B}^q}(f^n(z), f^n(w)) = 0 \quad \forall z, w \in \mathbb{B}^q.$$

Moreover, it follows from (1) that $\mathrm{step}_f(z)$ has a clear interpretation in terms of the canonical semi-model. Indeed, for all $z \in \mathbb{B}^q$,

$$\mathrm{step}_f(z) = \lim_{n \to \infty} k_{\mathbb{B}^q}(f^n(z), f^n(f(z))) = k_{\mathbb{B}^d}(\ell(z), \ell(f(z))) = k_{\mathbb{B}^d}(\ell(z), \tau(\ell(z))). \tag{2}$$

2.1   Sketch of the Proof of Theorem 2.1

Part I Let us first consider the case when f is injective [3]. Then it is easy to see that the direct limit $\Omega := \lim_{\longrightarrow}(\mathbb{B}^q, (f^{m-n})_{0 \leq n \leq m})$ admits a natural complex structure. That is, define Ω as the quotient of $\mathbb{B}^q \times \mathbb{N}$ with respect to the equivalence relation $(x, n) \simeq' (y, u)$ iff $f^{m-n}(x) = f^{m-u}(y)$ eventually. Define the family of canonical maps $(h_n: \mathbb{B}^q \to \Omega)_{n \in \mathbb{N}}$ by $h_n(x) := [(x, n)]$. Since f is injective, we have that h_n is injective for all $n \in \mathbb{N}$. We endow Ω with a complex structure using the mappings (h_n) as charts. The compatibility condition of the charts follows from

$$h_m \circ f^{m-n} = h_n, \quad 0 \leq n \leq m. \tag{3}$$

If we define $\psi \in \mathrm{aut}(\Omega)$ as $\psi([(x, n)]) := [(f(x), n)]$, then (Ω, h_0, ψ) is a semi-model for f, with an injective intertwining mapping h_0 (such a semi-model is simply called a *model*). Notice that from the universal property of the direct limit it immediately follows that, if (Λ, j, φ) is a semi-model for f (with no assumptions on Λ), then there exist a unique holomorphic map $\eta: \Omega \to \Lambda$ such that $j = \eta \circ h_0$ and $\eta \circ \psi = \varphi \circ \eta$, which is defined as

$$\eta([(x, n)]) := \varphi^{-n}(j(x)).$$

The problem with the model (Ω, h_0, ψ) is that the complex structure of Ω is not known in general. However, we know that Ω is the union of an increasing sequence of domains biholomorphic to $\mathbb{B}^q$:

$$\Omega = \bigcup_{n \in \mathbb{N}} h_n(\mathbb{B}^q). \tag{4}$$

Thus, by a result of Fornæss-Sibony [13], there exist an integer $0 \leq d \leq q$ and a surjective submersion $r: \Omega \to \mathbb{B}^d$ such that $k_\Omega = r^* k_{\mathbb{B}^d}$. In particular, two points $x, y \in \Omega$ belong to the same fiber of the map r if and only if $k_\Omega(x, y) = 0$. As a consequence the automorphism ψ of Ω induces an automorphism $\tau \in \mathrm{aut}(\mathbb{B}^d)$ such that $(\mathbb{B}^d, \ell := r \circ h_0, \tau)$ is a semi-model. It is easy to see that this is actually the canonical semi-model.

Part II Let us now consider the case when f is not necessarily injective [2]. In general there is no natural complex structure on the direct limit Ω, but we still can try to follow the previous argument as long as possible. Notice that in the injective case the base space $\mathbb{B}^d$ is the quotient of Ω with respect to the equivalence relation

$$[(x, n)] \simeq [(y, u)] \iff k_\Omega([(x, n)], [(y, u)]) = 0,$$

and that from (4) it follows that

$$k_\Omega([(x, n)], [(y, u)]) = \lim_{m \to \infty} k_{\mathbb{B}^q}(f^{m-n}(x), f^{m-u}(y)).$$

In the general case we thus define an equivalence relation $\simeq$ on the set Ω in the following way (notice that k_Ω is not defined since we did not endow Ω with a complex structure):

$$[(x,n)] \simeq [(y,u)] \in \Omega \iff \lim_{m\to\infty} k_{\mathbb{B}^q}(f^{m-n}(x), f^{m-u}(y)) = 0.$$

The bijective map $\psi : \Omega \to \Omega$ passes to the quotient, giving a bijective map $\widehat{\psi} : \Omega/_\simeq \to \Omega/_\simeq$ such that the following diagram commutes:

$$
\begin{array}{ccc}
\mathbb{B}^q & \xrightarrow{\ f\ } & \mathbb{B}^q \\
{\scriptstyle \pi_\simeq \circ h_0}\Big\downarrow & & \Big\downarrow{\scriptstyle \pi_\simeq \circ h_0} \\
\Omega/_\simeq & \xrightarrow[\ \widehat{\psi}\ ]{} & \Omega/_\simeq .
\end{array}
$$

The triple $(\Omega/_\simeq, \pi_\simeq \circ h_0, \widehat{\psi})$ is a good candidate for the canonical semi-model, as the following lemma shows.

Lemma 2.3 *If (Λ, j, φ) is a semi-model for f with Kobayashi hyperbolic base space, then the map $\eta : \Omega \to \Lambda$ given by the universal property of the direct limit passes to the quotient to a map $\widehat{\eta} : \Omega/_\simeq \to \Lambda$ such that the following diagram commutes.*

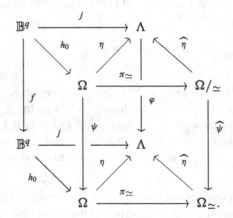

Proof It suffices to show that if $[(x,n)] \simeq [(y,u)] \in \Omega$, then $\eta([(x,n)]) = \eta([(y,u)])$. Let $m \geq n, u$. Then

$$k_\Lambda(\eta([(x,n)]), \eta([(y,u)])) = k_\Lambda(\eta([(f^{m-n}(x), m)]), \eta([(f^{m-u}(y), m)])) =$$

$$= k_\Lambda(\varphi^{-m} \circ j \circ f^{m-n}(x), \varphi^{-m} \circ j \circ f^{m-u}(y)) \leq k_{\mathbb{B}^q}(f^{m-n}(x), f^{m-u}(y)) \xrightarrow{m\to\infty} 0.$$

Since Λ is Kobayashi hyperbolic, this implies that $\eta([(x,n)]) = \eta([(y,u)])$. $\qquad\square$

We now show that, even if Ω does not admit a natural complex structure, its quotient $\Omega/_{\simeq}$ admits a natural complex structure which makes it biholomorphic to a ball $\mathbb{B}^d$, $0 \le d \le q$, and which makes the mappings $\pi_{\simeq} \circ h_0$, $\widehat{\psi}$ and $\widehat{\eta}$ holomorphic.

If (k_n) is a family of automorphisms of $\mathbb{B}^q$, we define for all $0 \le n \le m$

$$\tilde{f}_{n,m} := k_m \circ f^{m-n} \circ k_n^{-1}.$$

By choosing an appropriate family (k_n) we may assume that $\tilde{f}_{n,m}(0) = 0$ for all $0 \le n \le m$.

Remark 2.4 There exists a natural bijection between the direct limits $\Omega(f)$ and $\Omega(\tilde{f}_{n,m})$ which respects the equivalence relation $\simeq$.

Lemma 2.5 *There exists a holomorphic retract Z of $\mathbb{B}^q$ and a family $(\alpha_n \colon \mathbb{B}^q \to Z)$ such that*

$$\alpha_m \circ \tilde{f}_{n,m} = \alpha_n, \quad \forall\, 0 \le n \le m, \tag{5}$$

and such that the map $\gamma \colon \Omega(\tilde{f}_{n,m}) \to Z$ induced by the universal property of the direct limit passes to the quotient giving a natural bijection $\widehat{\gamma} \colon \Omega(\tilde{f}_{n,m})/_{\simeq} \to Z$. We can use $\widehat{\gamma}$ to pull back the complex structure from Z to $\Omega(\tilde{f}_{n,m})/_{\simeq}$.

Proof We show how to construct the holomorphic retract Z and the family $(\alpha_n \colon \mathbb{B}^q \to Z)$. We consider the sequence $(\tilde{f}_{0,m} \colon \mathbb{B}^q \to \mathbb{B}^q)$. Since $\tilde{f}_{0,m}(0) = 0$ there exists a subsequence

$$\tilde{f}_{0,m_0(h)} \overset{h \to \infty}{\longrightarrow} \alpha_0 \in \text{Hol}(\mathbb{B}^q, \mathbb{B}^q).$$

Now we extract a subsequence $m_1(h)$ from $m_0(h)$ such that

$$\tilde{f}_{1,m_1(h)} \overset{h \to \infty}{\longrightarrow} \alpha_1 \in \text{Hol}(\mathbb{B}^q, \mathbb{B}^q).$$

Iterating this procedure we obtain a family $(\alpha_n \colon \mathbb{B}^q \to \mathbb{B}^q)$ satisfying (5). Using a diagonal argument on the α_n's and taking a convergent subsequence we get a holomorphic retraction $\alpha \in \text{Hol}(\mathbb{B}^q, \mathbb{B}^q)$ which satisfies $\alpha \circ \alpha_n = \alpha_n$. But this implies that $\alpha_n(\mathbb{B}^q) \subset \alpha(\mathbb{B}^q) := Z$. For the proof that $\widehat{\gamma}$ is bijective, see [2]. $\square$

Notice that a holomorphic retract Z of $\mathbb{B}^q$ is biholomorphic to $\mathbb{B}^d$ for some $0 \le d \le q$. Thus the semi-model $(\Omega/_{\simeq}, \pi_{\simeq} \circ h_0, \widehat{\psi})$ is isomorphic to a semi-model of the form $(\mathbb{B}^d, \ell, \tau)$.

Part III We now show that the automorphism τ has the same dilation as f. It is proved in [3] that for any holomorphic self-map $g: \mathbb{B}^q \to \mathbb{B}^q$ the *divergence rate*

$$c(g) := \lim_{m \to \infty} \frac{k_{\mathbb{B}^q}(g^m(x), x)}{m} = \lim_{m \to \infty} \frac{\mathrm{step}_{g^m}(x)}{m}, \quad x \in \mathbb{B}^q$$

is well defined and, if f is not elliptic,

$$c(g) = -\log \lambda(g).$$

Thus

$$c(\tau) = \lim_{m \to \infty} \frac{k_{\mathbb{B}^d}(\ell(x), \tau^m(\ell(x)))}{m} = \lim_{m \to \infty} \frac{\mathrm{step}_{f^m}(x)}{m} = \lim_{m \to \infty} \frac{k_{\mathbb{B}^q}(f^m(x), x)}{m} = c(f),$$

where the second identity follows from (2).

If f is hyperbolic, then it follows that τ is hyperbolic with the same dilation and clearly $d > 0$. If f is parabolic nonzero step, then in order to prove that τ is a parabolic automorphism we just need to show that τ is not elliptic. But by (2) it follows that for all $w \in \ell(\mathbb{B}^q)$, $k_{\mathbb{B}^d}(w, \tau(w)) > 0$ (which also implies $d > 0$). It is easy to see that a similar argument works for every $w \in \mathbb{B}^d$. □

Remark 2.6 The proof of Theorem 2.1 can be dualized (inverting arrows and considering the inverse limit instead of the direct limit) to obtain results on pre-models and backward orbits, see [1, 2].

Definition 2.7 Let $f: \mathbb{B}^q \to \mathbb{B}^q$. The *type* of f is the dimension of its canonical semi-model.

3   Parabolic Dichotomy

What can we say about the canonical semi-model $(\mathbb{B}^d, \ell, \tau)$ if $f: \mathbb{B}^q \to \mathbb{B}^q$ is parabolic but is not of non-zero step? Using the divergence rate we can immediately say that the automorphism τ cannot be hyperbolic.

The next example shows that the canonical semi-model could be just a point.

Example 3.1 Let $\mathbb{H}_q$ denote the Siegel half-space

$$\{(z, w) \in \mathbb{C} \times \mathbb{C}^{q-1} : \mathrm{Im}(z) > \|w\|^2\},$$

which is biholomorphic to $\mathbb{B}^q$. Consider the holomorphic map defined by $f(z, w) := (z + i, w)$. It is easy to see that f is the restriction of an automorphism F of $\mathbb{C}^q$ which leaves $\mathbb{H}_q$ invariant. Since $\bigcup_{n \in \mathbb{N}} F^{-n}(\mathbb{H}_q) = \mathbb{C}^q$, we see that f admits a model $(\mathbb{C}^q, \iota, F)$, where $\iota: \mathbb{H}_q \to \mathbb{C}^q$ is the inclusion. The base space of the canonical semi-model can be obtained quotienting $\mathbb{C}^q$ by the equivalence relation $x \simeq y \iff k_{\mathbb{C}^q}(x, y) = 0$. Thus the canonical semi-model is a point.

The following question is open.

Question 3.2 Does the following parabolic dichotomy hold? If $f: \mathbb{B}^q \to \mathbb{B}^q$ is parabolic, then either f has nonzero step, or for all $z, w \in \mathbb{B}^q$ we have

$$\lim_{n \to \infty} k_{\mathbb{B}^q}(f^n(z), f^n(w)) = 0.$$

If $q = 1$ this is true by Remark 1.6.

This question can be reformulated in an equivalent way.

Question 3.3 If $f: \mathbb{B}^q \to \mathbb{B}^q$ is parabolic, is it true that the canonical semi-model of f is either parabolic or a point?

Remark 3.4 In the polydisc $\mathbb{D}^q$ the Denjoy-Wolff Theorem fails and thus the dilation of a holomorphic self-map $f: \mathbb{D}^q \to \mathbb{D}^q$ without inner fixed points cannot be defined as in the unit ball. However, using the divergence rate $c(f)$ one can define $\lambda(f) := e^{-c(f)}$, and thus one can say what it means for f to be elliptic, parabolic or hyperbolic. The theory of canonical semi-models developed in [2, 3] also applies in the polydisc or in any bounded homogeneous domain (see also [5] by Gumenyuk and the author for a study of canonical models in the polydisc).

The following example shows that in the polydisc $\mathbb{D}^q$ parabolic dichotomy does not hold.

Example 3.5 Consider the domain $\mathbb{D} \times \mathbb{H}$ in $\mathbb{C}^2$, which is biholomorphic to the polydisc $\mathbb{D}^2$, and the holomorphic self-map $f(z, w) := (z, w + i)$. The mapping f has no fixed points and clearly $c(f) = 0$, hence f is parabolic. A canonical semi-model is given by $(\mathbb{D}, \pi_1, \mathrm{id}_{\mathbb{D}})$, which is elliptic.

Taking inspiration from this example, we can link Question 3.2 to a purely geometric open question.

Question 3.6 Does there exist a domain $A \subset \mathbb{D} \times \mathbb{C}$ biholomorphic to $\mathbb{B}^2$ such that

a) A is invariant under the map $(z, w) \mapsto (z, w + i)$,
b) for every point in $\mathbb{D} \times \mathbb{C}$ there exists $n \in \mathbb{N}$ such that $(z, w + in) \in A$?

Remark 3.7 For example, every domain of the form

$$A := \{(z, w) : \mathrm{Im}\, w > g(z)\},$$

where $g: \mathbb{D} \to \mathbb{R}$, satisfies assumptions a) and b) in Question 3.6. We do not know if such a domain can be biholomorphic to $\mathbb{B}^2$.

If the answer to Question 3.6 is positive, then the map $f(z, w) := (z, w + i)$ gives a parabolic self-map of $\mathbb{B}^2$ with canonical semi-model given by $(\mathbb{D}, \pi_1, \mathrm{id}_{\mathbb{D}})$. If the answer is negative, then with the help of the following proposition we obtain a partial answer to Questions 3.2–3.3.

Proposition 3.8 *Let $f: \mathbb{B}^2 \to \mathbb{B}^2$ be an injective holomorphic parabolic self-map. If f admits a canonical semi-model of the form $(\mathbb{D}, \pi_1, \mathrm{id}_{\mathbb{D}})$, then there exists a domain as in Question 3.6 biholomorphic to $\mathbb{B}^2$.*

Proof The map f admits a model (Ω, h_0, ψ), where Ω is a 2-dimensional complex manifold, and we know by assumption that the quotient of Ω by its Kobayashi pseudodistance is $\mathbb{D}$. Since $\Omega = \bigcup_{n \in \mathbb{N}} \psi^{-n}(h_0(\mathbb{B}^2))$, the manifold Ω is the increasing union of domains biholomorphic to $\mathbb{B}^2$. Hence, by a result of Fornæss–Sibony [13], it is biholomorphic to $\mathbb{D} \times \mathbb{C}$. Up to such a biholomorphism, $\psi \in \mathrm{aut}(\mathbb{D} \times \mathbb{C})$ has the form $\psi(z, w) = (z, a(z)w + b(z))$, where $a: \mathbb{D} \to \mathbb{C}^*$ and $b: \mathbb{D} \to \mathbb{C}$ are holomorphic. Since f has no fixed points, the automorphism ψ cannot have fixed points, but this means that $a \equiv 1$. Hence, up to biholomorphic conjugacy, we have $\psi(z, w) = (z, w + i)$. The domain $h_0(\mathbb{B}^2) \subset \mathbb{D} \times \mathbb{C}$ is clearly biholomorphic to $\mathbb{B}^2$, it is invariant by ψ and satisfies $\bigcup_{n \in \mathbb{N}} \psi^{-n}(h_0(\mathbb{B}^2)) = \mathbb{D} \times \mathbb{C}$. $\qquad\square$

4 Commuting Mappings

Cowen [11] proved in 1984 that, in the unit disc $\mathbb{D}$, a hyperbolic holomorphic self-map f and a parabolic holomorphic self-map g cannot commute. There cannot be a direct generalization of this result in higher dimensions, as shown by the following example, studied by Ostapyuk [17] in the setting of boundary repelling fixed points.

Example 4.1 Define two self-maps of $\mathbb{H}_2$ as $f(z, w) := (2z + iw^2, w)$ and $g(z, w) := (z + i - 2w, w - i)$. It is easy to verify that f and g commute. The map g is a parabolic automorphism, while f is a hyperbolic map with canonical semi-model given by

$$(\mathbb{H}, (z, w) \mapsto z + iw^2, \zeta \mapsto 2\zeta).$$

In this example we have type $f = 1$, type $g = 2$. It turns out that in higher dimension there are obstructions to the commutation of an hyperbolic map f and a parabolic map g, but depending on the types of f and g. For clarity we will consider the case of $\mathbb{B}^2$, even if the results are valid in $\mathbb{B}^q$, see [4] by Bracci and the author.

Theorem 4.2 *Let $f: \mathbb{B}^2 \to \mathbb{B}^2$ be a hyperbolic holomorphic self-map and let $g: \mathbb{B}^2 \to \mathbb{B}^2$ be a parabolic holomorphic self-map. Recall that type $f \geq 1$. Then*

(1) *If type $f = 2$, then f and g cannot commute.*
(2) *If type $g = 0$, then f and g cannot commute.*
(3) *Assume type $f =$ type $g = 1$. If f and g commute, then g necessarily has canonical semi-model of the form $(\mathbb{D}, \pi_1, \mathrm{id}_{\mathbb{D}})$.*

Proof of (3) Since type $g = 1$, it admits a canonical semi-model of the form $(\mathbb{D}, h, \varphi)$, where the automorphism φ is not hyperbolic. By the universal property f

induces a unique holomorphic self-map $\eta\colon \mathbb{D} \to \mathbb{D}$ such that the following diagram commutes:

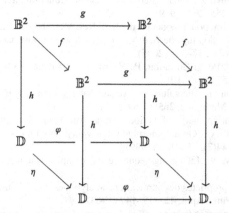

It follows from the construction of a simultaneous canonical model for f and g given in [4] that $c(\eta) = c(f)$, and thus η is a hyperbolic self-map commuting with φ. Thus by Cowen's theorem in the disc the automorphism φ is not parabolic. The only elliptic automorphism of the disc which commutes with a map without fixed points is the identity. $\qquad\square$

Remark 4.3

- Case (1) generalizes a result of de Fabritiis–Gentili [12] who considered the case when f is an automorphism.
- It is an open question whether two commuting maps f and g as in case (3) may exist. A positive answer would also show that parabolic dichotomy does not hold, answering Question 3.2.
- If we consider self-maps of the polydisc instead of the ball, then two maps f and g as in case (3) may actually commute: consider e.g. the holomorphic self-maps $f(z, w) := (2z, w + i)$ and $g(z, w) := (z, w + i)$ of $\mathbb{H} \times \mathbb{H}$.

Acknowledgements This work was supported by the SIR grant "NEWHOLITE—New methods in holomorphic iteration" n. RBSI14CFME.

References

1. Arosio, L.: The stable subset of a univalent self-map. Math. Z. **281**(3–4), 1089–1110 (2015)
2. Arosio, L.: Canonical models for the forward and backward iteration of holomorphic maps. J. Geom. Anal. **27**(2), 1178–1210 (2017)
3. Arosio, L., Bracci, F.: Canonical models for holomorphic iteration. Trans. Am. Math. Soc. **368**(5), 3305–3339 (2016)
4. Arosio, L., Bracci, F.: Simultaneous models for commuting holomorphic self-maps of the ball. Adv. Math. **321**, 486–512 (2017)

5. Arosio, L., Gumenyuk, P.: Valiron and Abel equations for holomorphic self-maps of the polydisc. Int. J. Math. **27**(4) (2016)
6. Baker, I.N., Pommerenke, C.: On the iteration of analytic functions in a half-plane II. J. Lond. Math. Soc. (2) **20**(2), 255–258 (1979)
7. Bayart, F.: The linear fractional model on the ball. Rev. Mat. Iberoam. **24**(3), 765–824 (2008)
8. Bracci, F., Gentili, G.: Solving the Schröder equation at the boundary in several variables. Michigan Math. J. **53**(2), 337–356 (2005)
9. Bracci, F., Gentili, G., Poggi-Corradini, P.: Valiron's construction in higher dimensions. Rev. Mat. Iberoam. **26**(1), 57–76 (2010)
10. Cowen, C.C.: Iteration and the solution of functional equations for functions analytic in the unit disk. Trans. Am. Math. Soc. **265**(1), 69–95 (1981)
11. Cowen, C.C.: Commuting analytic functions. Trans. Am. Math. Soc. **283**, 685–695 (1984)
12. de Fabritiis, C., Gentili, G.: On holomorphic maps which commute with hyperbolic automorphisms. Adv. Math. **144**(2), 119–136 (1999)
13. Fornæss, J.E., Sibony, N.: Increasing sequences of complex manifolds. Math. Ann. **255**(3), 351–360 (1981)
14. Hervé, M.: Quelques propriétés des applications analytiques d'une boule à m dimensions dans elle-même. J. Math. Pures Appl. **42**, 117–147 (1963)
15. Jury, T.: Valiron's theorem in the unit ball and spectra of composition operators. J. Math. Anal. Appl. **368**(2), 482–490 (2010)
16. Königs, G.: Recherches sur les intégrales de certaines équations fonctionnelles. Ann. Sci. École Norm. Sup. (3) **1**, 3–41 (1884)
17. Ostapyuk, O.: Backward iteration in the unit ball. Ill. J. Math. **55**(4), 1569–1602 (2011)
18. Pommerenke, C.: On the iteration of analytic functions in a half plane. J. Lond. Mat. Soc. (2) **19**(3), 439–447 (1979)
19. Rosay, J.P., Rudin, W.: Holomorphic maps from $\mathbb{C}^n$ to $\mathbb{C}^n$. Trans. Am. Math. Soc. **310**(1), 47–86 (1988)
20. Valiron, G.: Sur l'itération des fonctions holomorphes dans un demi-plan. Bull. Sci. Math. **47**, 105–128 (1931)

Jordan Structures in Bounded Symmetric Domains

Cho-Ho Chu

Abstract We discuss how Jordan algebraic structures arise from the geometry of bounded symmetric domains and their useful role in the study of holomorphic functions on these domains.

Keywords Symmetric domains · Jordan structures · Iteration theory

2010 Mathematics Subject Classification 32M15, 32H50, 17C65, 58B12, 58C10

1 Introduction

Since É. Cartan's seminal work, Lie theory has been an important tool in the study of Riemannian symmetric spaces and their classification. It was found relatively recently that Jordan algebras and Jordan triple systems can be used to describe a large class of symmetric spaces which is also accessible in infinite dimension.

In finite dimensions, it is well-known [8, 16] that the Hermitian symmetric spaces of non-compact type, which form a subclass of Riemannian symmetric spaces, can be realized as bounded symmetric domains in spaces of several complex variables, of which the open unit disc in $\mathbb{C}$ is the simplest example. The concept of bounded symmetric domains as well as their classification by É. Cartan [8] can be extended to infinite dimension in the following way.

Let D be a domain in a complex Banach space V, that is, D is a nonempty open connected set in V. We call D a *symmetric domain* if each point $a \in D$ admits a symmetry $s_a : D \longrightarrow D$ (which is necessarily unique). A *symmetry* s_a at a is defined to be a biholomorphic map such that s_a^2 is the identity map

C.-H. Chu (✉)
Queen Mary, University of London, London, UK
e-mail: c.chu@qmul.ac.uk

© Springer International Publishing AG, part of Springer Nature 2017   43
F. Bracci (ed.), *Geometric Function Theory in Higher Dimension*,
Springer INdAM Series 26, https://doi.org/10.1007/978-3-319-73126-1_4

and a is an isolated fixed point of s_a. A symmetric domain is said to be *irreducible* if it is not (biholomorphic to) a Cartesian product of symmetric domains. Cartan's classification can be very briefly described by saying that every (finite dimensional) irreducible bounded symmetric domain is biholomorphic to the open unit ball of one of the six types of complex vector spaces of matrices over $\mathbb{C}$ or the Cayley algebra $\mathbb{O}$. More details are given in Sect. 3. It follows that every (finite dimensional) bounded symmetric domain can be realized as the open unit ball of an ℓ^∞-sum of these complex spaces of matrices. In contemporary terminology, these spaces are examples of finite dimensional *JB*-triples*. An infinite dimensional extension of this realization is the following result due to Kaup [23].

Theorem 1.1 *Let D be a bounded symmetric domain in a complex Banach space. Then D is biholomorphic to the open unit ball of a JB*-triple.*

A precursor and a finite dimensional version of the above result is due to Loos [30, 38]. In the next section, we discuss the main ingredients in the proof of the above theorem and various ramifications. All Banach spaces in this article are over the complex field $\mathbb{C}$.

2 JB*-Triples

A JB*-triple is a complex Banach space equipped with a Jordan triple product. Given a bounded symmetric domain D, how does one construct the JB*-triple in Theorem 1.1? A crucial and fundamental device is the seminal result of H. Cartan [7], which states that the automorphism group Aut D of biholomorphic self-maps on a bounded domain $D \subset \mathbb{C}^n$ carries the structure of a Lie group, and its infinite dimensional generalization due to Upmeier [41] and Vigué [43]. If D is symmetric but possibly infinite dimensional, then the Lie group Aut D induces a JB*-triple structure on the tangent space $T_a D$ at a chosen point $a \in D$ such that D is biholomorphic to the open unit ball of $T_a D$.

Now we provide some details. First, for completeness, we recall that a mapping $f : D_1 \longrightarrow D_2$ between open sets D_i in complex Banach spaces V_i $(i = 1, 2)$ is called *holomorphic* if it has a (Fréchet) *derivative* $f'(a)$ at every point $a \in D$, where $f'(a) : V_1 \longrightarrow V_2$ is a bounded linear map satisfying

$$\lim_{z \to a} \frac{\|f(z) - f(a) - f'(a)(z - a)\|}{\|z - a\|} = 0.$$

The mapping f is called *biholomorphic* if it is bijective, holomorphic and its inverse f^{-1} is also holomorphic. The biholomorphic self-maps on a domain D form a group Aut D under function composition, called the *automorphism group* of D.

Let U be the open unit disc in $\mathbb{C}$ and let $H(D, U)$ denote the space of holomorphic maps from a *bounded* domain D in a complex Banach space V to U. Then D is endowed with the *Carathéodory metric*

$$d(z, w) = \sup \left\{ \tanh^{-1} \left| \frac{f(z) - f(w)}{1 - f(z)\overline{f(w)}} \right| : f \in H(D, U) \right\} \qquad (z, w \in D).$$

Fix a point $a \in D$. Let $r > 0$ so that the open ball $B(a, 4r) = \{z \in D : d(z, a) < 4r\}$ is contained in D. Define a metric ρ on $\mathrm{Aut}\, D$ by

$$\rho(f, g) = \sup\{d(f(z), g(z)) : z \in B(a, r/2)\} \qquad (f, g \in \mathrm{Aut}\, D).$$

Note that the metric ρ is defined by the ball $B(a, r/2)$ of radius $r/2$. When equipped with the topology induced by the metric ρ, the automorphism group $\mathrm{Aut}\, D$ becomes a real Lie group, possibly infinite dimensional, whose Lie algebra $\mathrm{aut}\, D$ consists of all complete holomorphic vector fields on D and the Lie product is the usual Lie brackets of vector fields [41, 43]. A *holomorphic vector field* X on $D \subset V$ is a holomorphic selection of a tangent vector at each point in D. Identifying each tangent space with the ambient Banach space V, we can, and will, view X as a holomorphic map $X : D \longrightarrow V$. A holomorphic vector field X gives rise to a flow of local biholomorphic transformations of D by elementary theory of differential equations: for every $z_0 \in D$, there is a holomorphic map $g : T \times \Omega \longrightarrow D$ defined on a product of convex domains $T \subset \mathbb{C}$ and $\Omega \subset D$, with $0 \in T$ and $z_0 \in \Omega$, such that

$$\frac{\partial g(t, z)}{\partial t} = X(g(t, z)), \quad g(0, z_0) = z_0.$$

The vector field X is called *complete* if T can be chosen to contain the whole real line $\mathbb{R}$ and $\Omega = D$, in which case we write

$$\exp X(z) := g(1, z) \qquad (z \in D).$$

The map $\exp X : D \longrightarrow D$ is biholomorphic and we have the exponential mapping $\exp : X \in \mathrm{aut}\, D \mapsto \exp X \in \mathrm{Aut}\, D$. The real Lie algebra $\mathrm{aut}\, D$ is a Banach Lie algebra in the norm

$$\|X\| = \sup\{\|X(z)\| : z \in B(a, r/2)\} \qquad (X \in \mathrm{aut}\, D).$$

Now let D be a bounded symmetric domain in a complex Banach space V and fix a point $a \in D$ as before. The symmetry $s_a \in \mathrm{Aut}\, D$ induces an involutive automorphism $\theta = Ad\, s_a : \mathrm{aut}\, D \longrightarrow \mathrm{aut}\, D$ via differentiation. The involution θ has eigenvalues ± 1 and induces an eigenspace decomposition

$$\mathrm{aut}\, D = \mathfrak{k} \oplus \mathfrak{p}$$

where $\mathfrak{k}$ is the 1-eigenspace and $\mathfrak{p}$ the (-1)-eigenspace of θ. The eigenspace $\mathfrak{k}$ is the Lie algebra of the isotropy group $K = \{g \in \operatorname{Aut} D : g(a) = a\}$ whereas the eigenspace $\mathfrak{p}$ is *real* linearly isomorphic to the tangent space V via the evaluation map

$$X \in \mathfrak{p} \mapsto X(a) \in V.$$

The real vector space $\mathfrak{p}$ admits a complex structure, which is a map

$$J : \mathfrak{p} \longrightarrow \mathfrak{p}$$

satisfying $J^2 = -id$ and is defined by $(JX)(a) = iX(a) \in V$. Moreover, we have $J\theta = \theta J$ and

$$[JX, Y](a) = i[X, Y](a) \qquad (X \in \mathfrak{p}, Y \in \operatorname{aut} D).$$

With the complex structure J, the eigenspace $\mathfrak{p}$ becomes a complex vector space and is complex linearly isomorphic to V. We can now construct a Jordan triple product on V via the Lie triple product.

Lemma 2.1 *Let D be a bounded symmetric domain in a complex Banach space V. Then V has the structure of a Jordan triple, that is, there is a triple product $\{\cdot, \cdot, \cdot\} : V^3 \longrightarrow V$, called a Jordan triple product, which is complex linear in the outer variables but conjugate linear in the middle variable, and satisfies the Jordan triple identity*

$$\{u, v, \{x, y, z\}\} = \{\{u, v, x\}, y, z\} - \{x, \{v, u, y\}, z\} + \{x, y, \{u, v, z\}\}$$

for all $u, v, x, y, z \in V$.

Proof Let $X \in \mathfrak{p} \mapsto X(a) \in V$ be the linear isomorphism given above. For $x, y, z \in V$ with $x = X(a), y = Y(a)$ and $z = Z(a)$, the desired triple product is defined by

$$\{x, y, z\} = -\frac{1}{4}[[X, Y], Z](a) + \frac{1}{4}[[JX, Y], JZ](a).$$

$\square$

The concept of a *Jordan triple* was introduced by Meyberg [35] in order to extend Koecher's construction [27, 28] of Lie algebras from Jordan algebras. This construction was also discovered independently by Kantor [20, 21] and Tits [40], which is now called the *Tits-Kantor-Koecher construction*. More details of the construction can be found in [9].

A real (complex) *Jordan algebra* [19, 31, 32] is a commutative, but not necessarily associative, algebra $\mathcal{A}$ over $\mathbb{R}$ (respectively, $\mathbb{C}$) satisfying the Jordan identity

$$(ab)a^2 = a(ba^2) \qquad (a, b \in \mathcal{A}).$$

A complex Jordan algebra $\mathcal{A}$ with an involution $*$ is a Jordan triple with the canonical Jordan triple product defined by

$$\{a, b, c\} = (ab^*)c + a(b^*c) - b^*(ac) \qquad (a, b, c \in \mathcal{A}).$$

Given a Jordan triple V and $z, w \in V$, one can define the *box operator* $z \,\square\, w : V \longrightarrow V$ by

$$z \,\square\, w(x) = \{z, w, x\} \qquad (x \in V).$$

In Lemma 2.1, the Banach space V, which is linearly isomorphic to the tangent space $T_a D$, together with the Jordan triple product $\{\cdot, \cdot, \cdot\}$, need not be a JB*-triple in its original norm. We need to renorm V to a JB*-triple.

Definition 2.2 A complex Banach space V is called a *JB*-triple* if it is a Jordan triple with a continuous Jordan triple product $\{\cdot, \cdot, \cdot\}$ satisfying

 (i) $z \,\square\, z$ is a Hermitian operator on V, that is, $\| \exp it(z \,\square\, z)\| = 1$ for all $t \in \mathbb{R}$;
 (ii) $z \,\square\, z$ has non-negative spectrum;
 (iii) $\|z \,\square\, z\| = \|z\|^2$

for all $z \in V$.

Example 2.3 Let $\mathcal{L}(H, K)$ be the Banach space of bounded linear operators between two Hilbert spaces H and K. Let T^* denote the adjoint of an operator $T \in \mathcal{L}(H, K)$. A closed subspace $V \subset \mathcal{L}(H, K)$, satisfying $AB^*C \in V$ whenever $A, B, C \in V$, is a JB*-triple, with the triple product

$$\{A, B, C\} = \frac{1}{2}(AB^*C + CB^*A).$$

In particular, Hilbert spaces and C*-algebras are JB*-triples.

Although JB*-triples are often viewed and studied as a generalization of C*-algebras, it cannot be overstated that the predominant role of JB*-triples is in complex geometry.

Example 2.4 The upper half complex plane has been generalised to Siegel domains in higher dimensions. The celebrated result of Koecher [28] and Vinberg [44] establishes the one-one correspondence between the (finite-dimensional) formally real Jordan algebras and Siegel *tube domains*, which are bounded symmetric domains. A real Jordan algebra $\mathcal{A}$ is called *formally real* if $a_1^2 + \cdots + a_n^2 = 0$ implies $a_1 = \cdots = a_n = 0$ for $a_1, \ldots, a_n \in \mathcal{A}$. The complexification $\mathcal{A}_{\mathbb{C}} = \mathcal{A} + i\mathcal{A}$ is a Jordan algebra with a natural involution $*$, and hence a Jordan triple with the canonical triple product. A finite dimensional formally real Jordan algebra $\mathcal{A}$ contains an identity and the set $\mathcal{I}_{\mathcal{A}}$ of invertible elements is non-empty. In finite dimensions, the cone

$$\mathcal{A}_+^0 = \{a^2 : a \in \mathcal{I}_{\mathcal{A}}\}$$

in a formally real Jordan algebra $\mathcal{A}$ carries the structure of a Riemannian symmetric space and $T_\mathcal{A} = \mathcal{A}_+^0 + i\mathcal{A}$ is the corresponding *tube domain* in the complex Jordan algebra $\mathcal{A}_\mathbb{C}$. In fact, $\mathcal{A}_\mathbb{C}$ is a JB*-triple in a suitable norm in which the tube domain $T_\mathcal{A}$ is biholomorphic to the open unit ball of $\mathcal{A}_\mathbb{C}$ via the Cayley transform (cf. [42]), as in the case of the upper half plane $T_\mathbb{R}$ which is biholomorphic to the open unit disc U.

The following lemma completes the proof of Theorem 1.1.

Lemma 2.5 *Let D be a bounded symmetric domain in a complex Banach space V. Then D is biholomorphic to the open unit ball of a JB*-triple $V' = (V, \| \cdot \|_a)$.*

Proof View V as the tangent space $T_a D$ at a chosen point $a \in D$, as before. By Lemma 2.1, V is a Jordan triple. Define the Carathéodory norm on V by

$$\|v\|_a = \sup\{|f'(a)(v)| : f \in H(D, U), f(a) = 0\} \qquad (v \in V). \qquad (2.1)$$

Then $(V, \| \cdot \|_a)$ is the desired JB*-triple.   □

Remark 2.6 If, in the preceding lemma, D happens to be the open unit ball of V, then by choosing $a = 0$, the norm $\| \cdot \|_a$ is the original norm and V itself is a JB*-triple.

To achieve the results stated above, a very important key is the fact that each vector field $X \in \mathfrak{p}$ is a polynomial vector field of degree 2 and has the form

$$X(z) = X(a) - \{z, X(a), z\} \qquad (z \in D). \qquad (2.2)$$

Complete proofs of what has been asserted previously can be found in the original work of Kaup [22, 23] (see also [9, 42]).

By identifying a bounded symmetric domain D as the open unit ball of a JB*-triple V, one can describe the geometry of D by the Jordan structures of V. This provides a useful and effective Jordan approach to geometric function theory on both finite and infinite dimensional bounded symmetric domains. Indeed, the ensuing discussion reveals that the Jordan triple product of V encodes the geometry of D.

We have already seen in (2.2) that the vector fields in $\mathfrak{p}$ are determined by the Jordan triple product and pointed out the significance of the box operator $z \,\square\, z$: $x \in V \mapsto \{z, z, x\} \in V$. There are two other fundamental mappings relating the geometry of D to Jordan structures, namely, the Bergman operator and the Möbius transformation, which are defined in terms of the Jordan triple product. For $a, b \in V$, the *Bergman operator* $B(a, b) : V \to V$ is defined by

$$B(a, b)(v) = v - 2\{a, b, v\} + \{b, \{a, v, a\}, b\} \qquad (v \in V).$$

Each $c \in D$ induces a biholomorphic map $g_c : D \to D$, called a *Möbius transformation*, given by

$$g_c(x) = c + B(c,c)^{1/2}(I + x \square c)^{-1}(x) \qquad (x \in D)$$

where I denotes the identity operator on V and $I + x \square c$ is invertible since $\|x \square c\| \le \|x\|\|c\| < 1$. We note that the Bergman operator $B(a,b)$ is linear and continuous. It is invertible if $\|a\|\|b\| < 1$ and for $\|a\| < 1$, the operator $B(a,a)$ has non-negative spectrum and hence the square roots $B(a,a)^{\pm 1/2}$ exist. The inverse of g_c is g_{-c} and the derivative of g_c is given by

$$g_c'(a) = B(c,c)^{1/2}B(a,-c)^{-1} \qquad (a \in D)$$

(cf. [23, (2.18)]). The automorphism group $\operatorname{Aut} D$ is completely determined by the Möbius transformations:

$$\operatorname{Aut} D = \{\varphi \circ g_c : c \in D, \varphi \text{ is a linear isometry on } V\}$$

(cf. [9, Proposition 3.2.6]), which acts transitively on D.

For a finite dimensional domain D, the Bergman metric h on D is given by

$$h_p(z,w) = \operatorname{Trace}(B(p,p)^{-1}(z) \square w) \qquad (p \in D, z, w \in V)$$

[30, Theorem 2.10].

The following converse of Theorem 1.1, shown in [23], can be deduced readily via Möbius transformations.

Theorem 2.7 *Let V be a JB*-triple. Then its open unit ball is a bounded symmetric domain.*

Proof Let D be the open unit ball of V. Evidently $s_0(z) = -z$ is a symmetry at the origin 0. To see that every point $a \in D$ has a symmetry, we only need to 'move' the symmetry s_0 to a via Möbius transformations. The symmetry s_a at a is given by $s_a = g_a \circ s_0 \circ g_a^{-1}$. □

3 Holomorphic Dynamics

One can find in literature many applications of Jordan structures in complex geometry, we discuss here a recent one in holomorphic dynamics [13], which serves as an example to illustrate their useful role in geometric function theory. An advantage of the Jordan approach is that it enables us to deal with infinite dimensional domains effectively. For instance, one can use it to extend function theory on the open unit disc $U \subset \mathbb{C}$ to bounded symmetric domains of any dimension.

We begin with the celebrated Denjoy-Wolff Theorem, which states that the iterates (f^n) of a fixed-point free holomorphic map $f : U \to U$ converge uniformly on compact sets to a constant function $h(\cdot) = \xi$ with $\xi \in \partial U$, where $f^n = f \circ \cdots \circ f$ (n-times), $\partial U = \overline{U} \backslash U$ and $\overline{U}$ denotes the closure of U.

The Denjoy-Wolff Theorem has been widely studied and generalized to higher dimensional domains, see for instance [1–3, 5, 6, 11, 33, 37] and the references therein. In view of the fact that U is, up to biholomorphic equivalence, the only one-dimensional bounded symmetric domain, a natural question arises: can one extend the Denjoy-Wolff Theorem to arbitrary bounded symmetric domains?

The answer is 'yes' for Euclidean balls

$$B_d = \{(z_1, \ldots, z_d) \in \mathbb{C}^d : |z_1|^2 + \cdots + |z_d|^2 < 1\}.$$

This result was first proved by Hervé [18]. However, a negative answer for the bidisc $U \times U$ may not come as a surprise. The iterates in this case need not converge to a constant function. Instead, Hervé [17] has shown that, given a fixed-point free holomorphic self-map f on $U \times U$, one of the following three conditions holds:

  (i) there exists $(\xi, \eta) \in \partial U \times \partial U$ such that (f^n) converges uniformly on compact sets to the constant function $h(\cdot, \cdot) = (\xi, \eta)$;
 (ii) there exists $\xi \in \partial U$ such that each subsequential limit $h = \lim_k f^{n_k}$, in the topology of uniform convergence on compact sets, takes values in $\{\xi\} \times \overline{U}$;
(iii) there exists $\eta \in \partial U$ such that each subsequential limit $h = \lim_k f^{n_k}$ takes values in $\overline{U} \times \{\eta\}$.

The topological boundary $\partial(U \times U)$ of the bidisc can be decomposed into a union of disjoint (*holomorphic*) *boundary components* and we observe that in the above list, the sets $\{(\xi, \eta)\}$, $\{\xi\} \times U$ and $U \times \{\eta\}$ are such components of $\partial(U \times U)$. Hence Hervé's result can be restated as follows.

Hervé's Theorem *Given a fixed-point free holomorphic self-map f on the bidisc $U \times U$, there is one single boundary component $K \subset \partial(U \times U)$ such that all subsequential limits $h = \lim_k f^{n_k}$ of (f^n) satisfy $h(U \times U) \subset \overline{K}$, where $\overline{K}$ is the closure of K.*

For the open unit disc U, the boundary components of ∂U are exactly the boundary points $\{\xi\}$ ($\xi \in \partial U$) and we can therefore view Hervé's theorem as a generalization of the Denjoy-Wolff Theorem. The question now is whether one can extend Hervé's theorem to all bounded symmetric domains. Although a complete answer seems not yet available at present, there are some recent results which feature Jordan theory and should be of interest. Before discussing the details, let us make precise the notion of a *boundary component* and recall some definitions.

Let D be a convex domain in a complex Banach space V. Following [26] (see also [30]), we call a subset C of the closure $\overline{D}$ a *boundary component* if the following conditions are satisfied:

(i) $C \neq \emptyset$;
(ii) for each holomorphic map $f : U \longrightarrow V$ with $f(U) \subset \overline{D}$, either $f(U) \subset C$ or $f(U) \subset \overline{D} \backslash C$;
(iii) C is minimal with respect to (i) and (ii).

Two boundary components are either equal or disjoint. The interior D is the unique open boundary component of $\overline{D}$, all others are contained in the topological boundary ∂D [26]. By a slight abuse of language, we also call the latter boundary components *of* ∂D. For each $a \in \overline{D}$, we denote by K_a the boundary component containing a.

Given a holomorphic map $h : D \to \overline{D}$, the image $h(D)$ is entirely contained in a single boundary component of $\overline{D}$ (cf. [10, Lemma 3.3]).

In infinite dimension, we need a stronger topology for convergence of holomorphic maps. Let D be a domain in a complex Banach space V and denote henceforth by $\overline{D}$ the closure of D and, by ∂D the topological boundary of D. A subset $B \subset D$ is said to be *strictly* contained in D if $\inf\{\|b - z\| : b \in B, z \in V \backslash D\} > 0$. Let $C(D, \overline{D})$ be the space of continuous functions from D to $\overline{D}$. We equip $C(D, \overline{D})$ with the *topology of locally uniform convergence* so that a sequence (f_n) of functions converges to f in this topology if and only if it converges uniformly on any open ball B strictly contained in D. Only in the finite dimensional case does this topology coincide with the topology of uniform convergence on compacta (cf. [15, Lemma IV.3.2]).

The subspace $H(D, \overline{D})$ of $C(D, \overline{D})$ consisting of holomorphic maps $f : D \to V$ with $h(D) \subset \overline{D}$ is closed in the topology of locally uniform convergence. Given a sequence (f_n) in $H(D, \overline{D})$, a function $h : D \to \overline{D}$ is called a *limit function* of the sequence (f_n) if there is a subsequence (f_{n_k}) of (f_n) converging to h in the topology of locally uniform convergence.

Given a fixed-point free holomorphic self-map f on an infinite dimensional domain D, a limit function h of the iterates (f^n) may have values entirely in D in which case the Dejoy-Wolff Theorem fails. This can happen to a *Hilbert ball*, which is the open unit ball of a Hilbert space V, if $\dim V = \infty$. In [39], an example of a fixed-point free biholomorphic self-map f on an infinite dimensional Hilbert ball D is given for which there is a limit function h of (f^n) such that $h(D) \subset D$. To ensure that all limit functions take values in the boundary ∂D of a domain D in a Banach space V, we impose a compactness condition on a holomorphic map $f : D \to D$. We say that f is *compact* if the closure $\overline{f(D)}$ is compact in V. If D is a finite dimensional *bounded* domain, then all holomorphic self-maps on D are automatically compact.

In the sequel, we will focus on compact holomorphic self-maps and make use of the result in [29] which implies that, given a fixed-point free *compact* holomorphic self-map f on the open unit ball B of a Banach space, all limit functions of the iterates (f^n) have values in the boundary ∂B.

Concerning the previous question of generalising the Denjoy-Wolff Theorem to higher dimensions, the following version has been proved recently in [13].

Theorem 3.1 *Let D be a finite-rank bounded symmetric domain in a complex Banach space V and let f : D → D be a fixed-point free compact holomorphic map. Then there is one single boundary component K in the boundary ∂D such that for each limit function h of (f^n), we have $h(D) \subset \overline{K}$ whenever h(D) is weakly closed in V.*

A finite dimensional bounded symmetric domain is of finite rank, but the converse is false. A full generalisation of the Denjoy-Wolff Theorem to all bounded symmetric domains would be achieved if one could remove from the above theorem the assumption that D is of finite rank and that h(D) is weakly closed. This problem appears to be open. There are some partial generalizations of Hervé's theorem to various bounded symmetric domains in current literature (e.g. [10, 12, 34]) which, however, fail to show that *all* limit functions of the iterates (f^n) accumulate in *only one single* boundary component of the boundary. Nevertheless, we discuss below several crucial stages in the proof of Theorem 3.1 which make substantial use of Jordan theory.

From now on, let D be a bounded symmetric domain realized as the open unit ball of a JB*-triple V. The *rank* of D can be defined in terms of the Jordan structures of V. A closed subspace E of a JB*-triple V is called a *subtriple* if $a, b, c \in E$ implies $\{a, b, c\} \in E$. For each $a \in V$, let $V(a)$ be the smallest closed subtriple of V containing a. For $V \neq \{0\}$, the *rank* of V is defined to be

$$r(V) = \sup\{\dim V(a) : a \in V\} \in \mathbb{N} \cup \{\infty\}.$$

The *rank* of D is defined to be $r(V)$. A (nonzero) JB*-triple V has *finite rank*, that is, $r(V) < \infty$ if, and only if, V is a reflexive Banach space (see [25, Proposition 3.2]). A C*-algebra is of finite rank if, and only if, it is finite dimensional.

A finite-rank JB*-triple can be 'coordinatized' by elements called *tripotents*. An element e in a JB*-triple is called a *tripotent* if $\{e, e, e\} = e$. In a C*-algebra, the tripotents are exactly the partial isometries. An extreme point ξ in $\overline{D}$ is a tripotent for which the Bergman operator $B(\xi, \xi)$ vanishes (see [9, Theorem 3.2.3; pp.32-33]). We note that an infinite-rank JB*-triple may not have any nonzero tripotent. For instance, the C*-algebra $C_0(\mathbb{R})$ of complex continuous functions on $\mathbb{R}$ vanishing at infinity has no nonzero tripotent.

A nonzero tripotent e is called *minimal* if $\{e, V, e\} = \mathbb{C}e$. Two elements $a, b \in V$ are said to be mutually (triple) *orthogonal* if $a \square b = b \square a = 0$, which is equivalent to $a \square b = 0$ [9, Lemma 1.2.32]. For a finite-rank JB*-triple V, its rank $r(V)$ is the (unique) cardinality of a maximal family of mutually orthogonal minimal tripotents and V is an ℓ^∞-sum of a finite number of finite-rank Cartan factors, which can be infinite dimensional. There are six types of finite-rank Cartan factors, listed below.

Type I $\mathcal{L}(\mathbb{C}^r, K)$ $(r = 1, 2, \ldots)$,

Type II $\{z \in \mathcal{L}(\mathbb{C}^r, \mathbb{C}^r) : z^t = -z\}$ $(r = 5, 6, \ldots)$,

Type III $\{z \in \mathcal{L}(\mathbb{C}^r, \mathbb{C}^r) : z^t = z\}$ $(r = 2, 3, \ldots)$,

Type IV   spin factor,

Type V $M_{1,2}(\mathbb{O}) = \{1 \times 2 \text{ matrices over the Cayley algebra } \mathbb{O}\}$,

Type VI $M_3(\mathbb{O}) = \{3 \times 3 \text{ Hermitian matrices over } \mathbb{O}\}$,

where $\mathcal{L}(\mathbb{C}^r, K)$ is the JB*-triple of linear operators from $\mathbb{C}^r$ to a Hilbert space K and z^t denotes the transpose of z in the JB*-triple $\mathcal{L}(\mathbb{C}^r, \mathbb{C}^r)$ of $r \times r$ complex matrices. A *spin factor* is a JB*-triple V equipped with a complete inner product $\langle \cdot, \cdot \rangle$ and a conjugation $* : V \to V$ satisfying

$$\langle x^*, y^* \rangle = \langle y, x \rangle \quad \text{and} \quad \{x, y, z\} = \frac{1}{2} \left(\langle x, y \rangle z + \langle z, y \rangle x - \langle x, z^* \rangle y^* \right).$$

The only possible infinite dimensional finite-rank Cartan factors are the spin factors and $\mathcal{L}(\mathbb{C}^r, K)$, with $\dim K = \infty$, where a spin factor has rank 2 and $\mathcal{L}(\mathbb{C}^r, K)$ has rank r. The open unit balls of the finite dimensional Cartan factors are exactly the six types of irreducible bounded symmetric domains in É. Cartan's classification. This explains the etymology of *Cartan factor*.

Besides the fundamental mappings mentioned before, another important Jordan tool in applications is the Peirce decomposition of a JB*-triple V, induced by a tripotent. Let $e \in V$ be a tripotent. Then there exists a *Peirce decomposition*

$$V = V_0(e) \oplus V_1(e) \oplus V_2(e)$$

where each $V_k(e)$, called the *Peirce k-space*, is an eigenspace

$$V_k(e) = \left\{ z \in V : (e \,\square\, e)(z) = \frac{k}{2} z \right\} \quad (k = 0, 1, 2)$$

of the operator $e \,\square\, e$, and is the range of the contractive projection $P_k(e) : V \longrightarrow V$ given by

$$P_0(e) = B(e, e); \quad P_1(e) = 4(e \,\square\, e - (e \,\square\, e)^2); \quad P_2(e) = 2(e \,\square\, e)^2 - e \,\square\, e.$$

We call $P_k(e)$ the *Peirce k-projection* and refer to [9, p. 32] for more detail. By [9, Corollary 1.2.46], a tripotent c is orthogonal to e if and only if $c \in V_0(e)$. The Peirce spaces $V_j(e)$ are subtriples of V. In a finite-rank domain D, the boundary component $K_e \subset \partial D$ containing a tripotent e is given by $K_e = e + V_0(e) \cap D$.

Let $\{e_1, \ldots, e_n\}$ be a family of mutually orthogonal tripotents in a JB*-triple V. For $i, j \in \{0, 1, \ldots, n\}$, the *joint Peirce space* V_{ij} is defined by

$$V_{ij} := \{z \in V : 2\{e_k, e_k, z\} = (\delta_{ik} + \delta_{jk})z \text{ for } k = 1, \ldots, n\},$$

where δ_{ij} is the Kronecker delta and $V_{ij} = V_{ji}$.

The decomposition

$$V = \bigoplus_{0 \leq i \leq j \leq n} V_{ij}$$

is called a *joint Peirce decomposition* (cf. [30]).

Now we are ready to explain the proof of Theorem 3.1 which uses Jordan theory. Following the one-dimensional approach to the Denjoy-Wolff Theorem, we first establish the existence of invariant domains of a given fixed-point free holomorphic map $f : D \longrightarrow D$, which generalizes Wolff's theorem for the disc $U \subset \mathbb{C}$ and is of independent interest. This result, stated below, is achieved by computing various norms of operators involving the Möbius transformation and the Bergman operator. In particular, we make use of the norm estimates

$$\|B(a,b)\| \leq (1 + \|a\|\|b\|)^2, \qquad \|B(a,a)^{-1/2}\| = \frac{1}{1 - \|a\|^2}$$

where the first inequality follows from direct computation and the second identity has been derived in [24] (see also [9, Proposition 3.2.13].) A complete proof of the following theorem is given in [13], in which some analysis of the structures of JB*-triples is also needed to show that the invariant domains are non-empty.

Theorem 3.2 *Let f be a fixed-point free compact holomorphic self-map on a bounded symmetric domain D. Then there is a sequence (z_k) in D converging to a boundary point $\xi \in \overline{D}$ such that, for each $\lambda > 0$, the set*

$$H(\xi, \lambda) = \left\{ x \in D : \limsup_{k \to \infty} \frac{1 - \|z_k\|^2}{1 - \|g_{-z_k}(x)\|^2} < \frac{1}{\lambda} \right\}$$

is a non-empty convex domain and f-invariant, i.e. $f(H(\xi, \lambda)) \subset H(\xi, \lambda)$. Moreover, $D = \bigcup_{\lambda > 0} H(\xi, \lambda)$ and $0 \in \bigcap_{\lambda < 1} H(\xi, \lambda)$.

This result generalizes Wolff's theorem to bounded symmetric domains completely. Indeed, if D is the open unit disc $U \subset \mathbb{C}$, the invariant domain $H(\xi, \lambda)$ is given by

$$H(\xi, \lambda) = t_\lambda^2 \xi + (1 - t_\lambda^2)U, \qquad t_\lambda^2 = \lambda/(1 + \lambda)$$

which is a horodisc in U, with closure internally tangent to U at the horocentre ξ and consequently, the Denjoy-Wolff Theorem follows readily by letting λ tend to infinity since the horodiscs $H(\xi, \lambda)$ then shrink to ξ and all limit functions of the iterates (f^n) take values in the boundary ∂U. This phenomenon extends to Hilbert balls provided f is compact (cf. [11]). Without compactness, we have already noted that a limit function of (f^n) need not have any value in the boundary although in this case, $H(\xi, \lambda)$ still shrinks to ξ as $\lambda \to \infty$. In fact, instead of compactness, one has the following necessary and sufficient conditions for the Denjoy-Wolff Theorem for

Hilbert balls, as shown in [12, Corollary 3.6]. These conditions are satisfied by a compact map f.

Proposition 3.3 *Let f be a fixed-point free holomorphic self-map on a Hilbert ball D. The following conditions are equivalent.*

  (i) *(f^n) converges locally uniformly to a constant map taking value in ∂D.*
 (ii) *$\lim_{n\to\infty} \|f^{2n}(a)\| = 1$ for some $a \in D$.*
(iii) *One orbit $(f^n(a))$ converges for some $a \in D$.*

It is of interest to compare the above result with the following observation.

Proposition 3.4 *Let f be a fixed-point free holomorphic self-map on a bounded symmetric domain D such that one orbit $(f^n(a))$ converges for some $a \in D$. Then all limit functions of (f^n) take values in one single boundary component K in $\overline{D}$.*

Proof Let h and k be limit functions of (f^n). Then, as noted before, $h(D)$ is contained in some boundary component $K_1 \subset \overline{D}$ and $k(D)$ is contained in another boundary component, say $K_2 \subset \overline{D}$. By assumption, we have $h(a) = k(a) \in K_1 \cap K_2$ and hence $K_1 = K_2$. □

The previous discussion reveals that, to make use of the invariant domains $H(\xi, \lambda)$ in the study of iterates of a holomorphic map f on bounded symmetric domains other than Hilbert balls, we would need a more explicit description of $H(\xi, \lambda)$. In the case of the unit disc U, they coincide with the horodiscs, which makes the computation simple and transparent. Therefore our next task is to describe the domains $H(\xi, \lambda)$ more explicitly as some kind of *horoballs*. We use Jordan structures to do so.

In the unit disc U, the horodisc $H(\xi, \lambda) = t_\lambda^2 \xi + (1 - t_\lambda^2)U$ can be formulated in terms of the Bergmann operator $B(a, b) : z \in \mathbb{C} \mapsto (1 - a\overline{b})^2 z \in \mathbb{C}$. Indeed we have

$$H(\xi, \lambda) = t_\lambda^2 \xi + B(t_\lambda \xi, t_\lambda \xi)^{1/2} U.$$

It is this Jordan description of $H(\xi, \lambda)$ which we aim to generalise to other bounded symmetric domains. We are able to do so for finite-rank domains.

Given a fixed-point free holomorphic self-map f on the unit disc U, the f-invariant horodisc $H(\xi, \lambda)$ can be constructed as the limit of a sequence of Poincaré discs, which are open discs in U defined by the Poincaré metric. To extend this construction to infinite dimension, we define a *horoball* in a bounded symmetric domain D as a limit of Kobayashi balls, which are open balls in D defined by the Kobayashi distance. The Kobayashi distance generalizes the Poincaré distance and has a Jordan description. Given $x, y \in D$, the Kobayashi distance between x and y is given by

$$\kappa(x, y) = \tanh^{-1} \|g_{-x}(y)\| .$$

where g_{-x} is the Möbius transformation induced by $-x$. The Kobayashi distance κ is a metric on D.

Now let $f : D \to D$ be a fixed-point free compact holomorphic map on a bounded symmetric domain D. Generalizing Wolff's one-dimensional horodisc, we now define a *horoball* using the map f as follows. Alternative descriptions of horospheres in various domains have been given in [1, 4, 36].

Choose an increasing sequence (α_k) in $(0, 1)$ with limit 1. Then $\alpha_k f$ maps D strictly inside itself and by the fixed-point theorem of Earle and Hamilton [14], we have $\alpha_k f(z_k) = z_k$ for some $z_k \in D$. Note that $z_k \neq 0$. Since $f(D)$ is relatively compact, we may assume, by choosing a subsequence if necessary, that (z_k) converges to a point $\xi \in \overline{D}$. Since f has no fixed point in D, the point ξ must lie in the boundary ∂D.

Given $\lambda > 0$, pick a sequence (r_k) in $(0, 1)$ such that

$$1 - r_k^2 = \lambda(1 - \|z_k\|^2)$$

from some k onwards. For each r_k, define a *Kobayashi ball*, centred at z_k, by

$$D_k(\lambda) = \{z \in D : \kappa(z, z_k) < \tanh^{-1} r_k\} = \{x \in D : \|g_{-z_k}(x)\| < r_k\}$$

which generalises the one-dimensional Poincaré disc. We note that

$$D_k(\lambda) = g_{z_k}(D(0, r_k)) \quad \text{where} \quad D(0, r_k) = \{z \in V : \|z\| < r_k\}. \tag{3.3}$$

For $x \in \overline{D}$ and $0 < r < 1$, using (3.3) and the formula

$$g_z(rx) = (1 - r^2)B(rz, rz)^{-1/2}(z) + rB(z, z)^{1/2}B(rz, rz)^{-1/2}g_{rz}(x), \tag{3.4}$$

as in [33, Proposition 2.3], one can write

$$D_k(\lambda) = (1 - r_k^2)B(r_k z_k, r_k z_k)^{-1/2}(z_k) + r_k B(z_k, z_k)^{1/2}B(r_k z_k, r_k z_k)^{-1/2}(D). \tag{3.5}$$

In particular, $D_k(\lambda)$ is a convex domain.

Now we define the *limit* of the sequence $(D_k(\lambda))_k$ of Kobayashi balls as the set

$$S(\xi, \lambda) = \{x \in \overline{D} : x = \lim_k x_k, \, x_k \in D_k(\lambda)\} \tag{3.6}$$

which contains ξ since $z_k \in D_k(\lambda)$, and is also convex because each $D_k(\lambda)$ is convex. We call $S(\xi, \lambda)$ a *closed horoball* at ξ and its interior $S_0(\xi, \lambda)$ a *horoball* at ξ. The latter is contained in D.

For finite-rank D, the following result reveals that $H(\xi, \lambda) = S_0(\xi, \lambda)$, which resembles Wolff's horodisc in U.

Theorem 3.5 *Let f be a fixed-point free compact holomorphic self-map on a bounded symmetric domain D of finite rank p. Then there is a sequence (z_k) in*

D converging to a boundary point

$$\xi = \sum_{j=1}^{m} \alpha_j e_j \qquad (\alpha_j > 0, \ m \in \{1, \ldots, p\})$$

where $e_1, \ldots, e_m$ are orthogonal minimal tripotents in ∂D, such that for each $\lambda > 0$, the convex f-invariant domain $H(\xi, \lambda)$ is the horoball $S_0(\xi, \lambda)$ at ξ, which has the form

$$S_0(\xi, \lambda) = \sum_{j=1}^{m} \frac{\sigma_j \lambda}{1 + \sigma_j \lambda} e_j + B \left(\sum_{j=1}^{m} \sqrt{\frac{\sigma_j \lambda}{1 + \sigma_j \lambda}} e_j, \ \sum_{j=1}^{m} \sqrt{\frac{\sigma_j \lambda}{1 + \sigma_j \lambda}} e_j \right)^{1/2} (D)$$

(3.7)

and is affinely homeomorphic to D, where $\sigma_j \geq 0$ and $\max\{\sigma_j : j = 1, \ldots, m\} = 1$.

A complete proof of this result has been given in [13]. It requires a detailed analysis of the Peirce decomposition of finite-rank JB*-triples as well as computations involving the Bergmann operators and Möbius transformations. If D is a Hilbert ball, which is of rank $p = 1$, the horoball reduces to

$$S_0(\xi, \lambda) = \frac{\lambda}{1 + \lambda} \xi + B \left(\sqrt{\frac{\lambda}{1 + \lambda}} \xi, \ \sqrt{\frac{\lambda}{1 + \lambda}} \xi \right)^{1/2} (D)$$

and if D is the one-dimensional unit disc U, it is the horodisc

$$S_0(\xi, \lambda) = \frac{\lambda}{1 + \lambda} \xi + \frac{1}{1 + \lambda} U.$$

Finally, using the f-invariance of the horoballs $S_0(\xi, \lambda)$ and some limiting arguments, one can show that the range $h(D)$ of each limit function h of the iterates (f^n) is contained in the closure $\overline{K_e} \subset \partial D$ of the boundary component K_e containing a tripotent $e \in \partial D$, if $h(D)$ is weakly closed. Here the tripotent e is given by $e = e_1 + \cdots + e_m$, where $e_1, \ldots, e_m$ are the mutually orthogonal minimal tripotents in Theorem 3.5. This proves Theorem 3.1.

4 Example

We conclude with an instructive example. Although it is known that the Denjoy-Wolff Theorem could fail for non-compact holomorphic self-maps on a Hilbert ball [39], the following example reveals that compactness is not a necessary condition

for a Denjoy-Wolff type theorem (see also [12]). This example also shows that the
image $h(D)$ of a limit function h can be a singleton or a whole boundary component
which is neither closed nor open.

Let D be a bounded symmetric domain of finite rank p, realised as the open unit
ball of a JB*-triple V of rank p. Pick any nonzero $a \in D$. Then there exist mutually
orthogonal minimal tripotents $e_1, \ldots, e_p$ in V and scalars $\alpha_1, \ldots, \alpha_p \geq 0$ such that

$$a = \alpha_1 e_1 + \cdots + \alpha_p e_p \quad (\|a\| = \alpha_1 \geq \cdots \geq \alpha_p \geq 0)$$

which is called a *spectral decomposition* of a.

Let $g_a : D \to D$ be the Möbius transformation induced by a, which is not a
compact map if D is infinite dimensional. Let $x = \beta_1 e_1 + \beta_2 e_2 + \cdots + \beta_p e_p$, where
$\beta_1, \beta_2, \ldots, \beta_p \in U$ so that $x \in D$. By orthogonality, we have

$$x \,\square\, a = (\beta_1 e_1 + \beta_2 e_2 + \cdots + \beta_p e_p) \,\square\, (\alpha_1 e_1 + \cdots + \alpha_p e_p) = \beta_1 \alpha_1 e_1 \,\square\, e_1 + \cdots + \beta_p \alpha_p e_p \,\square\, e_p$$

and

$$(x \,\square\, a)^n (x) = \beta_1^{n+1} \alpha_1^n e_1 + \cdots + \beta_p^{n+1} \alpha_p^n e_p \qquad (n = 1, 2, \ldots).$$

It follows that

$$
\begin{aligned}
g_a(x) &= a + B(a,a)^{1/2}(I + x \,\square\, a)^{-1}(x) \\
&= a + B(a,a)^{1/2}(I - x \,\square\, a + (x \,\square\, a)^2 - (x \,\square\, a)^3 + \cdots)(x) \\
&= a + B(a,a)^{1/2}(\beta_1 e_1 + \beta_2 e_2 + \cdots + \beta_p e_p - (\beta_1^2 \alpha_1 e_1 + \cdots + \beta_p^2 \alpha_p e_p) + \cdots) \\
&= a + B(a,a)^{1/2}[(1 - \beta_1 \alpha_1 + \beta_1^2 \alpha_1^2 + \cdots)\beta_1 e_1 + \cdots \\
&\qquad + (1 - \beta_p \alpha_p + \beta_p^2 \alpha_p^2 + \cdots)\beta_p e_p)] \\
&= a + B(a,a)^{1/2}\left(\frac{\beta_1 e_1}{1 + \beta_1 \alpha_1} + \cdots + \frac{\beta_p e_p}{1 + \beta_p \alpha_p}\right) \\
&= \alpha_1 e_1 + \cdots + \alpha_p e_p + \frac{(1 - \alpha_1^2)\beta_1 e_1}{1 + \beta_1 \alpha_1} + \cdots + \frac{(1 - \alpha_p^2)\beta_p e_p}{1 + \beta_p \alpha_p} \\
&= \frac{\alpha_1 + \beta_1}{1 + \alpha_1 \beta_1} e_1 + \cdots + \frac{\alpha_p + \beta_p}{1 + \alpha_p \beta_p} e_p \\
&= g_{\alpha_1}(\beta_1)e_1 + \cdots + g_{\alpha_p}(\beta_p)e_p
\end{aligned}
$$

where g_{α_j} is the Möbius transformation on the complex disc U, induced by α_j for
$j = 1, \ldots, p$. If $\alpha_j = 0$, then g_{α_j} is the identity map. If $\alpha_j > 0$, then the iterates $(g_{\alpha_j}^n)$
converge locally uniformly to the constant map with value $\alpha_j / |\alpha_j| = 1$. Hence the

iterates

$$g_a^n(x) = g_{\alpha_1}^n(\beta_1)e_1 + \cdots + g_{\alpha_p}^n(\beta_p)e_p \qquad (n = 2, 3, \ldots)$$

converge to

$$e_1 + \gamma_2 e_2 + \cdots + \gamma_p e_p, \quad \gamma_j = \begin{cases} 1 & (\alpha_j > 0) \\ \beta_j & (\alpha_j = 0) \end{cases} \qquad (j = 2, \ldots, p).$$

In particular, if $\alpha_j > 0$ for all j, then the iterates (g_a^n) converge locally uniformly to a constant map with value $\xi = e_1 + \cdots + e_p \in \partial D$. On the other hand, if $J = \{j : \alpha_j > 0\}$ is a proper subset of $\{1, \ldots, p\}$, then

$$\lim_n g_a^n(x) = \sum_{j \in J} e_j + \sum_{j \notin J} \beta_j e_j \in e + D_e$$

where $e = \sum_{j \in J} e_j$ is a tripotent in ∂D and $D_e = V_0(e) \cap D$. It follows that, in this case, the image of every limit function h of (g_a^n) is the whole boundary component $K_e = e + D_e$ which is neither closed nor open.

References

1. Abate, M.: Horospheres and iterates of holomorphic maps. Math. Z. **198**, 225–238 (1988)
2. Abate, M., Raissy, J.: Wolff-Denjoy theorems in non-smooth convex domains. Ann. Mat. Pura Appl. **193**, 1503–1518 (2014)
3. Arosio, L., Bracci, F.: Canonical models for holomorphic iteration. Trans. Am. Math. Soc. **368**, 3305–3339 (2016)
4. Bassanelli, G.: On horospheres and holomorphic endomorfisms of the Siegel disc. Rend. Sen. Mat. Univ. Padova **70**, 147–165 (1983)
5. Budzynska, M.: The Denjoy-Wolff theorem in $\mathbb{C}^n$. Nonlinear Anal. **75**, 22–29 (2012)
6. Budzyńska, M., Kuczumow, T., Reich, S.: Theorems of Denjoy-Wolff type. Ann. Mat. Pura Appl. **192**, 621–648 (2013)
7. Cartan, H.: Les fonctions de deux variables complexes et le problème de la représentation analytique. J. Math. Pures et Appl. **10**, 1–114 (1931)
8. Cartan, É.: Sur les domaines bornés homogènes de l'espace de n variables complexes. Abh. Math. Semin. Univ. Hamburg **11**, 116–162 (1935)
9. Chu, C.-H.: Jordan Structures in Geometry and Analysis. Cambridge Tracts in Mathematics, vol. 190. Cambridge University Press, Cambridge (2012)
10. Chu, C.-H.: Iteration of holomorphic maps on Lie balls. Adv. Math. **264**, 114–154 (2014)
11. Chu, C.-H., Mellon, P.: Iteration of compact holomorphic maps on a Hilbert ball. Proc. Am. Math. Soc. **125**, 1771–1777 (1997)
12. Chu, C.-H., Rigby, M.: Iteration of self-maps on a product of Hilbert balls. J. Math. Anal. Appl. **411**, 773–786 (2014)
13. Chu, C.-H., Rigby, M.: Horoballs and iteration of holomorphic maps on bounded symmetric domains. Adv. Math. **311**, 338–377 (2017)
14. Earle, C.J., Hamilton, R.S.: A fixed point theorem for holomorphic mappings. Proc. Symp. Pure Math. **16**, 61–65 (1969)

15. Franzoni, T., Vesentini, E.: Holomorphic Maps and Invariant Distance. Mathematics Studies, vol. 40. North-Holland, Amsterdam (1980)
16. Harish-Chandra: Representations of semi-simple Lie groups VI. Am. J. Math. **78**, 564–628 (1956)
17. Hervé, M.: Iteration des transformations analytiques dans le bicercle-unité. Ann. Sci. Ecole Norm. Sup. **71**, 1–28 (1954)
18. Hervé, M.: Quelques propriétés des applications analytiques d'une boule a m dimensions dans elle-meme. J. Math. Pures et Appl. **42**, 117–147 (1963)
19. Jacobson, N.: Structure and representations of Jordan algebras. Amer. Math. Soc. Colloq. Publ. **39** (1968)
20. Kantor, I.L.: Classification of irreducible transitive differential groups. Dokl. Akad. Nauk SSSR **158**, 1271–1274 (1964)
21. Kantor, I.L.: Transitive differential groups and invariant connections on homogeneous spaces. Trudy Sem. Vecktor. Tenzor. Anal. **13**, 310–398 (1966)
22. Kaup, W.: Algebraic characterization of symmetric complex Banach manifolds. Math. Ann. **228**, 39–64 (1977)
23. Kaup, W.: A Riemann mapping theorem for bounded symmetric domains in complex Banach spaces. Math. Z. **183**, 503–529 (1983)
24. Kaup, W.: Hermitian Jordan triple systems and their automorphisms of bounded symmetric domains. In: González, S. (ed.) Non-associative Algebra and Its Applications, Oviedo, 1993, pp. 204–214. Kluwer, Dordrecht (1994)
25. Kaup, W.: On a Schwarz lemma for bounded symmetric domains. Math. Nachr. **197**, 51–60 (1999)
26. Kaup, W., Sauter, J.: Boundary structure of bounded symmetric domains. Manuscripta Math. **101**, 351–360 (2000)
27. Koecher, M.: Imbedding of Jordan algebras into Lie algebras I. Bull. Am. J. Math. **89**, 787–816 (1967)
28. Koecher, M.: Jordan Algebras and Their Applications. University of Minnesota, 1962, Lecture Notes in Mathematics, vol. 1710. Springer, Heidelberg (1999)
29. Kapeluszny, J., Kuczmow, T., Reich, S.: The Denjoy-Wolff theorem in the open unit ball of a strictly convex Banach space. Adv. Math. **143**, 111–123 (1999)
30. Loos, O.: Bounded symmetric domains and Jordan pairs (Mathematical Lectures). University of California, Irvine (1977)
31. McCrimmon, K.: Jordan algebras and their applications. Bull. Am. Math. Soc. **84**, 612–627 (1978)
32. McCrimmon, K.: A Taste of Jordan Algebras, Universitext. Springer, Heidelberg (2004)
33. Mellon, P.: Holomorphic invariance on bounded symmetric domains. J. Reine Angew. Math. **523**, 199–223 (2000)
34. Mellon, P.: Dejoy-Wolff theory for finite-dimensional bounded symmetric domains. Ann. Mate. **195**, 845–855 (2016)
35. Meyberg, K.: Jordan-Tripelsysteme und die Koecher-Konstruktion von Lie-Algebren. Math. Z. **115**, 58–78 (1970)
36. Nirenberg, L., Webster, S., Yang, P.: Local boundary regularity of holomorphic mappings. Comm. Pure Appl. Math. **33**, 305–338 (1980)
37. Reich, S., Shoikhet, D.: The Denjoy-Wolff Theorem. In: Proceedings of Workshop on Fixed Point Theory (Kazimierz Dolny, 1997). Ann. Univ. Mariae Curie-Sk lodowska Sect. A51, pp. 219–240 (1997)
38. Satake, I.: Algebraic Structures of Symmetric Domains. Princeton University Press, Princeton (1980)
39. Stachura, A.: Iterates of holomorphic self-maps on the unit ball of a Hilbert space. Proc. Am. Math. Soc. **93**, 88–90 (1985)
40. Tits, J.: Une classe d'algèbres de Lie en relation avec les algèbres de Jordan. Indag. Math. **24**, 530–535 (1962)

41. Upmeier, H.: Über die Automorphismengruppe von Banach-Mannigfaltigkeiten mit invarianter Metrik. Math. Ann. **223**, 279–288 (1976)
42. Upmeier, H.: Jordan Algebras in Analysis, Operator Theory and Quantum Mechanics. CBMS. American Mathematical Society, Providence (1987)
43. Vigué, J.P.: Le groupe des automorphismes analytiques d'un domaine borné d'un espace de Banach complexe. Application aux domaines bornés symétriques. Ann. Sc. Ec. Norm. Sup. **9**, 203–282 (1976)
44. Vinberg, E.B.: The theory of convex homogeneous cones. Trudy Moskov. Mat. Obsc. **12**, 303–358 (1963)

On Runge Neighborhoods of Closures of Domains Biholomorphic to a Ball

Hervé Gaussier and Cezar Joiţa

Abstract We give an example of a domain W in $\mathbb{C}^3$, biholomorphic to a ball, such that W is not Runge in any Stein neighborhood of $\overline{W}$.

Keywords Approximation of univalent maps · Runge domain · Fatou-Bieberbach domains

Mathematics Subject Classification 32E30, 30E10

1 Introduction

During the conference "Geometric Function Theory in Higher Dimension", held in Cortona in September 2016, Filippo Bracci asked the following question: suppose that W is a domain in $\mathbb{C}^n$, biholomorphic to a ball. Does there exist a Fatou-Bieberbach domain U such that $W \subset U \subset\subset \mathbb{C}^n$ and W is Runge in U?

A related (and natural) question is the following: suppose that W is a domain in $\mathbb{C}^n$ which is biholomorphic to a ball. Does there exist a Stein domain U such that $\overline{W} \subset U \subset \mathbb{C}^n$ and W is Runge in U? The purpose of this note is to show that the answer to the second question is negative, by constructing a counter-example. As it can be seen from our construction, this does not answer Filippo Bracci's question since we construct a domain W with a "bad" point x in the boundary ∂W and, according to the statement of our problem, this point must be in U.

H. Gaussier
University of Grenoble Alpes, CNRS, Grenoble, France
e-mail: herve.gaussier@univ-grenoble-alpes.fr

C. Joiţa (✉)
Simion Stoilow Institute of Mathematics of the Romanian Academy, Bucharest, Romania
e-mail: Cezar.Joita@imar.ro

© Springer International Publishing AG, part of Springer Nature 2017
F. Bracci (ed.), *Geometric Function Theory in Higher Dimension*,
Springer INdAM Series 26, https://doi.org/10.1007/978-3-319-73126-1_5

63

For the basic notions regarding pseudoconvexity we refer, for example, to [1]. For a complex manifold M we denote by $\mathcal{O}(M)$ the ring of holomorphic functions and, if K is a compact subset of M, $\widehat{K}^M$ stands for the holomorphically convex hull of K in M, $\widehat{K}^M = \{x \in M : |f(x)| \le \|f\|_K, \ \forall f \in \mathcal{O}(M)\}$. If M is a Stein manifold and D is a Stein open subset of M, then D is called Runge in M if the restriction map $\mathcal{O}(M) \to \mathcal{O}(D)$ has a dense image. This is equivalent to the fact that for every compact set $K \subset D$ we have $\widehat{K}^M = \widehat{K}^D$. It is also a standard fact that if M is a Stein manifold, D is a Stein open subset of M which is Runge in M and N is a closed complex submanifold of M, then $N \cap D$ is Runge in N.

2 The Example

The following map was defined by J. Wermer, [2] and [3]: $f : \mathbb{C}^3 \to \mathbb{C}^3, f(z, w, t) = (z, zw + t - 1, zw^2 - w + 2wt)$. A direct computation shows that $f_{|\mathbb{C} \times \mathbb{C} \times \{t : \mathrm{Re}(t) < \frac{1}{2}\}}$ is a biholomorphism onto its image.

Let $0 < p < \frac{1}{4}$ be a fixed real number and let

$$B := \left\{ (z, w, t) \in \mathbb{C}^3 : p|z|^2 + p|w|^2 + |t|^2 < \frac{1}{4} \right\}.$$

Then B is biholomorphic to a ball and $B \subset \mathbb{C} \times \mathbb{C} \times \{t : |t| < \frac{1}{2}\}$. Hence $f(B)$ is also biholomorphic to a ball. We would like to show that $f(B)$ is the required example.

Suppose that U is a pseudoconvex neighborhood of $\overline{f(B)}$. Note that $\left(0, -\frac{1}{2}, 0 \right) = f\left(0, 0, \frac{1}{2} \right) \in \partial f(B)$. Hence, for a sufficiently small $r > 0$, we have that

$$\left\{ (\xi, \eta, \theta) \in \mathbb{C}^3 : |\xi| \le r, |\eta + \frac{1}{2}| \le r, |\theta| \le r \right\} \subset U.$$

We make the following claim:

Claim If r is small enough, there exists $\alpha \in \mathbb{R}$, $-\frac{1}{2} < \alpha < -\frac{1}{2} + r$, such that

$$\{\xi \in \mathbb{C} : |\xi| = r\} \times \{\alpha\} \times \{0\} \subset f(B).$$

Proof of the Claim Note that $f\left(\xi, \frac{2\alpha + 1}{\xi}, -\alpha \right) = (\xi, \alpha, 0)$, for every $\xi \in \mathbb{C} \backslash \{0\}$, $\alpha \in \mathbb{R}$. Hence it suffices to show that if $r > 0$ is sufficiently small, there exists $\alpha \in \mathbb{R}$, $-\frac{1}{2} < \alpha < -\frac{1}{2} + r$, such that $\left(\xi, \frac{2\alpha + 1}{\xi}, -\alpha \right) \in B$ for every ξ with $|\xi| = r$.

In other words, we would like to show that if $r > 0$ is small enough, there exists $\alpha \in \mathbb{R}$, $-\frac{1}{2} < \alpha < -\frac{1}{2} + r$, such that

$$pr^2 + p\left(\frac{(2\alpha + 1)^2}{r^2}\right) + \alpha^2 < \frac{1}{4},$$

or:

$$g_r(\alpha) := \left(\frac{4p}{r^2} + 1\right)\alpha^2 + \frac{4p}{r^2}\alpha + pr^2 + \frac{p}{r^2} < \frac{1}{4}.$$

Let $\alpha_0 := -\frac{2p}{4p + r^2} > -\frac{2p}{4p} = -\frac{1}{2}$. We have:

$$\alpha_0 < -\frac{1}{2} + r \iff \frac{r^2}{4p + r^2} < 2r \iff r < 2(4p + r^2),$$

which is obviously true for r small enough.
Moreover:

$$g_r(\alpha_0) = \frac{4p + r^2}{r^2}\frac{4p^2}{(4p + r^2)^2} - \frac{8p^2}{r^2(4p + r^2)} + \frac{pr^4 + p}{r^2}.$$

Hence:

$$g_r(\alpha_0) < \frac{1}{4} \iff \frac{-4p^2 + (4p + r^2)(pr^4 + p)}{r^2(4p + r^2)} < \frac{1}{4}$$

$$\iff pr^6 + 4p^2r^4 + pr^2 < pr^2 + \frac{r^4}{4}$$

$$\iff pr^2 < \frac{1}{4} - 4p^2.$$

Since $p < \frac{1}{4}$, we have that $\frac{1}{4} - 4p^2 > 0$ and therefore the last inequality is true for r small enough. Hence our claim is proved. $\square$

We remark now that $(0, \alpha, 0) \notin f(B)$ for $\alpha > -\frac{1}{2}$. Indeed

$$(0, \alpha, 0) = f(z, w, t) \iff (z = w = 0, \ t = 1 + \alpha), \text{ and then } t^2 > \frac{1}{4}.$$

To summarize, we found $r > 0$, sufficiently small, and α such that:

$$
\begin{cases}
\{\xi \in \mathbb{C} : |\xi| \le r\} \times \{\alpha\} \times \{0\} \subset U, \\[2mm]
\{\xi \in \mathbb{C} : |\xi| = r\} \times \{\alpha\} \times \{0\} \subset f(B), \\[2mm]
\qquad\qquad (0, \alpha, 0) \notin f(B).
\end{cases}
$$

This shows that $(\mathbb{C} \times \{\alpha\} \times \{0\}) \cap f(B)$ is not Runge in $(\mathbb{C} \times \{\alpha\} \times \{0\}) \cap U$, and therefore $f(B)$ is not Runge in U.

Remarks

1. The above example is relatively compact in $\mathbb{C}^3$. One can construct an unbounded example as follows: again we let p be a fixed positive real number, $p < \frac{1}{4}$, and we set:

$$
S = \left\{ (z, w, t) \in \mathbb{C}^3 : \mathrm{Re}(t) < -p|z|^2 - p|w|^2 + \frac{1}{2} \right\}.
$$

Then S is unbounded, biholomorphic to a ball, and $S \subset \mathbb{C} \times \mathbb{C} \times \{t \in \mathbb{C} : \mathrm{Re}(t) < \frac{1}{2}\}$. A completely similar argument shows that if U is a Stein neighborhood of $\overline{f(S)}$, then $f(S)$ is not Runge in U.

   It may be possible to construct a counter-example in $\mathbb{C}^2$, using the same procedure, following a construction of J. Wermer in [4].

2. The following interesting question was raised by the referee. Is there a natural number k such that a biholomorphic image of the ball which is C^k smooth has a Stein neighbourhood basis in which the domain is Runge?

Acknowledgements The article was discussed during the conference "Geometric Function Theory in Higher Dimension" held in Cortona, September 2016. The authors would like to thank the organizers for their invitation. The authors would also like to thank the referee for his useful comments. The author Hervé Gaussier was partially supported by ERC ALKAGE. The author Cezar Joiţa was partially supported by CNCS grant PN-III-P4-ID-PCE-2016-0330.

References

1. Hörmander, L.: An Introduction to Complex Analysis in Several Variables. D. Van Nostrand, Princeton, NJ (1966)
2. Wermer, J.: An example concerning polynomial convexity. Math. Ann. **139**, 147–150 (1959)
3. Wermer, J.: Addendum to "An example concerning polynomial convexity". Math. Ann. **140**, 322–323 (1960)
4. Wermer, J.: On a domain equivalent to the bidisk. Math. Ann. **248**, 193–194 (1980)

Parametric Representations and Boundary Fixed Points of Univalent Self-Maps of the Unit Disk

Pavel Gumenyuk

Abstract A classical result in the theory of Loewner's parametric representation states that the semigroup $\mathfrak{U}_*$ of all conformal self-maps ϕ of the unit disk $\mathbb{D}$ normalized by $\phi(0) = 0$ and $\phi'(0) > 0$ can be obtained as the reachable set of the Loewner–Kufarev control system

$$\frac{dw_t}{dt} = G_t \circ w_t, \quad t \geqslant 0, \qquad w_0 = \mathrm{id}_{\mathbb{D}},$$

where the control functions $t \mapsto G_t \in \mathsf{Hol}(\mathbb{D}, \mathbb{C})$ form a certain convex cone. Here we extend this result to the semigroup $\mathfrak{U}[F]$ consisting of all conformal $\phi : \mathbb{D} \to \mathbb{D}$ whose set of boundary regular fixed points contains a given finite set $F \subset \partial\mathbb{D}$ and to its subsemigroup $\mathfrak{U}_\tau[F]$ formed by $\mathrm{id}_{\mathbb{D}}$ and all $\phi \in \mathfrak{U}[F] \setminus \{\mathrm{id}_{\mathbb{D}}\}$ with the prescribed boundary Denjoy–Wolff point $\tau \in \partial\mathbb{D} \setminus F$. This completes the study launched in Gumenyuk, P. (Constr. Approx. 46 (2017), 435–458, https://doi.org/10.1007/s00365-017-9376-4), where the case of interior Denjoy–Wolff point $\tau \in \mathbb{D}$ was considered.

Keywords Parametric representation · Univalent function · Conformal mapping · Boundary fixed point · Loewner equation · Loewner-Kufarev equation · Infinitesimal generator · Evolution family · Lie semigroup

2010 Mathematics Subject Classification Primary: 30C35, 30C75; Secondary: 30D05, 30C80, 34H05, 37C25

P. Gumenyuk (✉)
Department of Mathematics and Natural Sciences, University of Stavanger, Stavanger, Norway
e-mail: pavel.gumenyuk@uis.no

© Springer International Publishing AG, part of Springer Nature 2017
F. Bracci (ed.), *Geometric Function Theory in Higher Dimension*,
Springer INdAM Series 26, https://doi.org/10.1007/978-3-319-73126-1_6

67

1 Introduction

One of the cornerstone results of Loewner's parametric representation can be stated
in the following form, see, e.g. [36, Problem 3 on p. 164] or [2, pp. 69–70].

Theorem A *A function ϕ defined in the unit disk $\mathbb{D} := \{z\colon |z| < 1\}$ is a univalent
(i.e. injective and holomorphic) self-map of $\mathbb{D}$ normalized by $\phi(0) = 0$, $\phi'(0) > 0$,
if and only if it can be represented in the form $\phi(z) = w_z(T)$ for all $z \in \mathbb{D}$, where
$T \geqslant 0$ and $w = w_z(t)$ is the unique solution to initial value problem*

$$\frac{dw(t)}{dt} = -w(t)\, p\big(w(t), t\big), \quad t \geqslant 0; \qquad w(0) = z, \tag{1.1}$$

*with a function $p : \mathbb{D} \times [0, +\infty) \to \mathbb{C}$ holomorphic in the first argument, measurable
in the second argument and satisfying $\operatorname{Re} p(\cdot, t) > 0$ and $p(0, t) = 1$ for a.e. $t \geqslant 0$.*

A related result, see, e.g. [36, §6.1], stating that *the class $\mathcal{S}$ of all univalent
mappings $f : \mathbb{D} \to \mathbb{C}$ normalized by $f(0) = f'(1) - 1 = 0$ coincides with the
set of all functions representable as*

$$f(z) = \lim_{t \to +\infty} e^t w_z(t),$$

where w_z, as above, is the solution to (1.1)*,* is much better known and usually meant
when one talks about *Loewner's parametric representation* of univalent functions.
However, it is the former version of Loewner's parametric method, i.e. Theorem A,
which will play more important role in the frames of the present study.

There is no natural linear structure compatible with the property of injectivity.
In fact, the class $\mathcal{S}$ mentioned above even fails to be a convex subset of $\operatorname{Hol}(\mathbb{D}, \mathbb{C})$.
A similar statement holds for the class $\mathfrak{U}_*$ of all univalent self-maps $\phi : \mathbb{D} \to \mathbb{D}$,
$\phi(0) = 0$, $\phi'(0) > 0$, involved in Theorem A.

That is why in many problems for univalent functions, standard variation
and optimization methods cannot be applied directly. One has to use a suitable
parametric representation to replace univalent mappings by a "parameter", such as
the driving function p in (1.1), ranging in a convex cone. In this way, the parametric
method has been used a lot in the study of the class $\mathcal{S}$, see, e.g. [2], [10, Sect. 6], and
references therein. We also mention recent studies [30, 39, 40], which make essential
use of Theorem A and its analogue for hydrodynamically normalized univalent self-
maps of the upper half-plane. Finally, it is worth mentioning that the univalence
comes to de Branges' proof of Bieberbach's famous conjecture [18] solely via a
slight modification of Theorem A.

In contrast to linear operations, the operation of composition preserves univalence. With this fact being one of the cornerstones of Loewner's parametric method,
univalent self-maps of $\mathbb{D}$ come to the scene. So let $\mathfrak{S}$ be a set of univalent maps
$\phi : \mathbb{D} \to \mathbb{D}$ containing $\operatorname{id}_\mathbb{D}$ and closed w.r.t. composition, i.e. satisfying $\psi \circ \phi \in \mathfrak{S}$

whenever $\phi, \psi \in \mathfrak{S}$. The following *heuristic scheme* of parametric representation of $\mathfrak{S}$ goes back to Loewner [31–34].

Consider the set $T\mathfrak{S}$ of all holomorphic vector fields $G : \mathbb{D} \to \mathbb{C}$ satisfying the following two conditions: (i) for any $z \in \mathbb{D}$ the initial value problem $dw/dt = G(w(t))$, $w(0) = z$ has a unique solution $t \mapsto w_z(t) \in \mathbb{D}$ defined for all $t \geqslant 0$, and (ii) for any $T \geqslant 0$, the time-T map $z \mapsto w_z(T)$, associated with the above ODE, belongs to $\mathfrak{S}$. Such holomorphic vector fields G can be also characterized as infinitesimal generators of one-parameter semigroups in $\mathfrak{S}$, see Sect. 3.2 for details.

Suppose that for each family $[0, +\infty) \ni t \mapsto G_t \in T\mathfrak{S}$ chosen from a certain class $\mathcal{M}$, the *non-autonomous* ODE $dw/dt = G_t(w)$ has, for any $z \in \mathbb{D}$, a unique solution $[0, +\infty) \ni t \mapsto w_z(t) \in \mathbb{D}$ satisfying the initial condition $w_z(0) = z$ and that the corresponding time-T map $z \mapsto w_z(T)$ belongs to $\mathfrak{S}$ for any $T \geqslant 0$. Denote by $\mathfrak{S}^L(\mathcal{M})$ the set of all time-T maps obtained in this way (when neither $T \geqslant 0$ nor the family $(G_t) \in \mathcal{M}$ is fixed).

To establish a *Loewner-type parametric representation* of $\mathfrak{S}$ means to find a class $\mathcal{M}_\mathfrak{S}$ of families (G_t) such that $\mathfrak{S}^L(\mathcal{M}_\mathfrak{S}) = \mathfrak{S}$, which would mean that any element of $\mathfrak{S}$ can be represented as the time-T map of $dw/dt = G_t(w)$ with suitable $(G_t) \in \mathcal{M}_\mathfrak{S}$ and $T \geqslant 0$. Note that, in general, existence of such a class $\mathcal{M}_\mathfrak{S}$ is not guaranteed.

This approach to parametric representation of univalent self-maps has been systematically developed in [21, 23–28]. In all known examples, including Theorem A, which provides a parametric representation of $\mathfrak{U}_*$, $\mathcal{M}_\mathfrak{S}$ is a convex set in the linear space of all maps from $[0, +\infty)$ to $\mathsf{Hol}(\mathbb{D}, \mathbb{C})$. This turns out to be important for applications. By this reason, in Definition 3.9 we additionally require that $\mathcal{M}_\mathfrak{S}$ is a convex cone.

For the study of holomorphic self-maps $\phi : \mathbb{D} \to \mathbb{D}$, in particular from the view point of dynamics, an important role is played by the so-called boundary (regular) fixed points, see, e.g. [3, 7, 13–15, 17, 20, 22, 35, 38], just to mention some studies on this topic. Combining Loewner's scheme of parametric representation with the intrinsic approach in Loewner Theory proposed in [8, 9], we will establish Loewner-type parametric representation for semigroups of univalent self-maps $\phi : \mathbb{D} \to \mathbb{D}$ defined by prescribing a finite set of points on $\partial\mathbb{D}$ which are fixed by ϕ.

To conclude the Introduction, let us mention that Loewner's scheme makes perfect sense also for univalent self-maps of the unit ball and polydisk in $\mathbb{C}^n$. However, Loewner-type parametric representations for univalent functions in $\mathbb{D}$ make essential use of the Riemann Mapping Theorem, which is not available in several complex variables. By this reason, even for simplest choices of $\mathfrak{S}$, e.g., for the semigroup consisting of all univalent self-maps ϕ of the unit ball $\mathbb{B}^n := \{z \in \mathbb{C}^n : \|z\| < 1\}$, $n > 1$, satisfying $\phi(0) = 0$, to describe the reachable set $\mathfrak{S}^L(\mathcal{M})$ is a hard open problem. It seems that the only known necessary condition for ϕ to be an element of $\mathfrak{S}^L(\mathcal{M})$ is that the domain $\phi(\mathbb{B}^n)$ has to be Runge in $\mathbb{C}^n$, see [4, Section 4].

2 Main Results

Denote by $\mathfrak{U}$ the semigroup formed by all univalent holomorphic mappings $\phi : \mathbb{D} \to \mathbb{D}$. Further, for $\tau \in \overline{\mathbb{D}}$, by $\mathfrak{U}_\tau$ we denote the subsemigroup of $\mathfrak{U}$ formed by $\mathrm{id}_\mathbb{D}$ and all $\phi \in \mathfrak{U} \setminus \{\mathrm{id}_\mathbb{D}\}$ whose Denjoy–Wolff point coincides with τ, see Definition 3.3. The main result of this paper is the analogue of the classical Theorem A for two families of subsemigroups in $\mathfrak{U}$ and in $\mathfrak{U}_\tau$ defined below.

Given a finite set $F \subset \partial\mathbb{D}$, by $\mathfrak{U}[F]$ we denote the subsemigroup of $\mathfrak{U}$ consisting of all $\phi \in \mathfrak{U}$ for which every element of F is a boundary regular fixed point, see Definitions 3.1 and 3.2. Let $\mathfrak{U}_\tau[F] := \mathfrak{U}_\tau \cap \mathfrak{U}[F]$ for any $\tau \in \overline{\mathbb{D}} \setminus F$.

Note that the semigroups defined above are always non-trivial, i.e. different from $\{\mathrm{id}_\mathbb{D}\}$, see, e.g. [29, Example 3.1].

Theorem 1 *For any finite set $F \subset \partial\mathbb{D}$ and any $\tau \in \overline{\mathbb{D}}\setminus F$, the semigroups $\mathfrak{S} = \mathfrak{U}[F]$ and $\mathfrak{S} = \mathfrak{U}_\tau[F]$ admit Loewner-type parametric representation, i.e. there exists a convex cone $\mathcal{M}_\mathfrak{S}$ of Herglotz vector fields in $\mathbb{D}$ with the following properties:*

(i) *for every $G \in \mathcal{M}_\mathfrak{S}$ and a.e. $s \geqslant 0$, $G(\cdot, s)$ is the infinitesimal generator of a one-parameter semigroup in $\mathfrak{S}$;*
(ii) *for every $\phi \in \mathfrak{S}$ there exists $G \in \mathcal{M}_\mathfrak{S}$ such that $\phi(z) = w_{z,0}^G(1)$ for all $z \in \mathbb{D}$, where $w = w_{z,s}^G(t)$ is the unique solution to the initial value problem*

$$\frac{dw(t)}{dt} = G(w(t), t), \quad w(s) = z, \qquad z \in \mathbb{D}, \quad t \geqslant s \geqslant 0;$$

(iii) *for every $G \in \mathcal{M}_\mathfrak{S}$, the mappings $z \mapsto w_{z,s}^G(t)$, $t \geqslant s \geqslant 0$, belong to $\mathfrak{S}$.*

Definitions and basic theory regarding Herglotz vector fields, infinitesimal generators, and one-parameter semigroups can be found in Sect. 3.

Remark 2.1 The convex cone $\mathcal{M}_\mathfrak{S}$ in Theorem 1 can be characterized explicitly. For $\mathfrak{S} = \mathfrak{U}[F]$, it coincides with the set of all Herglotz vector fields G such that:

(a) G satisfies condition (i) in Theorem 1, which, due to [15, Theorem 1], is equivalent to the existence of a finite angular limit $\lambda(\sigma, s) := \angle \lim_{z \to \sigma} G(z, s)/(z - \sigma)$ for a.e. $s \geqslant 0$ and every $\sigma \in F$;
(b) the functions $\lambda(\sigma, \cdot)$, $\sigma \in F$, are all locally integrable on $[0, +\infty)$.

The case $\mathfrak{S} = \mathfrak{U}_\tau[F]$ is similar. The only difference is that in (a), in addition to having a finite angular limit $\lambda(\sigma, s)$ for a.e. $s \geqslant 0$ and every $\sigma \in F$, we should require that for a.e. $s \geqslant 0$, $G(\cdot, s)$ is the infinitesimal generator of a one-parameter semigroup $(\phi_t^{G(\cdot, s)})$ for which τ is the Denjoy–Wolff point.

For $\mathfrak{U}_0[\{1\}]$, Theorem 1 was proved by Goryainov [26]. For $\mathfrak{U}_\tau[F]$ with an arbitrary finite set $F \subset \partial\mathbb{D}$ and $\tau \in \mathbb{D}$, it has been proved by the author in [29]. Moreover, the case of $\mathfrak{U}[F]$ with $\mathsf{Card}(F) \leqslant 3$ and that of $\mathfrak{U}_\tau[F]$ with $\mathsf{Card}(F) \leqslant 2$ and $\tau \in \partial\mathbb{D}$ are also covered in [29]. The proof in the remaining cases is presented in this paper. It is based on the following theorem.

Theorem 2 *Let* $F \in \partial \mathbb{D}$ *be any finite set with* $\mathsf{Card}(F) \geqslant 3$ *and let* $\tau \in \partial \mathbb{D} \setminus F$. *Then for any* $\phi \in \mathfrak{U}_\tau[F]$ *there exists an evolution family* $(\varphi_{s,t}) \subset \mathfrak{U}_\tau[F]$ *such that* $\phi = \varphi_{0,1}$.

The proofs of Theorems 1 and 2 are given in Sect. 6. Sections 4 and 5 contain some lemmas and other auxiliary statements, and in the next section we recall some basic definitions and facts from Holomorphic Dynamics and Loewner Theory in the unit disk used throughout the paper.

3 Preliminaries

3.1 Contact and Fixed Points of Holomorphic Self-Maps

Let $\phi \in \mathsf{Hol}(\mathbb{D}, \mathbb{D})$. By the Schwarz Lemma, ϕ can have at most one fixed point in $\mathbb{D}$. However, in the sense of angular limits, there can be many (boundary) fixed points on $\partial \mathbb{D}$.

Definition 3.1 A point $\sigma \in \partial \mathbb{D}$ is said to be a *boundary fixed point* of ϕ if the angular limit $\angle \lim_{z \to \sigma} \phi(z)$ exists and coincides with σ. More generally, $\sigma \in \partial \mathbb{D}$ is a *contact point* of ϕ if it is a boundary fixed point of $e^{i\theta} \phi$ for a suitable $\theta \in \mathbb{R}$.

Following a usual convention, if $\angle \lim_{z \to \xi} \phi(z)$ exists for some ξ, it will be denoted, in what follows, by $\phi(\xi)$ and $\phi'(\xi)$ will stand for the angular derivative of ϕ at ξ, i.e. $\phi'(\xi) := \angle \lim_{z \to \xi} (\phi(z) - \phi(\xi))/(z - \xi)$, again provided that the latter limit exists.

It is known that for any contact point σ of a holomorphic self-map $\phi : \mathbb{D} \to \mathbb{D}$, the angular derivative $\phi'(\sigma)$ exists, finite or infinite, and if it is finite, then $\sigma \phi'(\sigma)/\phi(\sigma) > 0$; in particular, $\phi'(\sigma) > 0$ or $\phi'(\sigma) = \infty$ at any boundary fixed point σ of ϕ. See, e.g. [1, Theorem 1.2.7 on p. 53] or [37, Proposition 4.13 on p. 82].

Definition 3.2 A boundary fixed point (or a contact point) σ of a holomorphic self-map $\phi : \mathbb{D} \to \mathbb{D}$ is said to be *regular* if the angular derivative $\phi'(\sigma)$ is finite.

Among all fixed points of a self-map $\phi \in \mathsf{Hol}(\mathbb{D}, \mathbb{D}) \setminus \{\mathrm{id}_\mathbb{D}\}$ there is one point of special importance for dynamics. Namely, by the classical Denjoy–Wolff Theorem, see, e.g. [1, Theorem 1.2.14, Corollary 1.2.16, Theorem 1.3.9], ϕ has a unique (boundary or interior) fixed point $\tau \in \overline{\mathbb{D}}$ such that $|\phi'(\tau)| \leqslant 1$. Moreover, the sequence of iterates $(\phi^{\circ n})$, $\phi^{\circ 1} := \phi$, $\phi^{\circ(n+1)} := \phi \circ \phi^{\circ n}$, converges (to the constant function equal) to τ locally uniformly in $\mathbb{D}$, unless ϕ is an elliptic automorphism.

Definition 3.3 The point τ above is referred to as the *Denjoy–Wolff point* of ϕ.

Remark 3.4 A version of the *Chain Rule for angular derivatives* states that if σ is a regular contact point of $\phi \in \mathsf{Hol}(\mathbb{D}, \mathbb{D})$ and $\omega := \phi(\sigma)$ is a contact point of $\psi \in \mathsf{Hol}(\mathbb{D}, \mathbb{D})$, then σ is a contact point of $\psi \circ \phi$ and $(\psi \circ \phi)'(\sigma) = \psi'(\omega) \cdot \phi'(\sigma)$; in particular, σ is a regular contact point of $\psi \circ \phi$ provided σ and ω are regular for ϕ

and ψ, respectively. See, e.g. [1, Lemma 1.3.25 on p. 92]. It follows that the classes of univalent holomorphic self-maps defined in the Introduction, $\mathfrak{U}[F]$ and $\mathfrak{U}_\tau[F]$, are semigroups w.r.t. the operation $(\psi, \phi) \mapsto \psi \circ \phi$.

3.2   One-Parameter Semigroups in $\mathsf{Hol}(\mathbb{D}, \mathbb{D})$

An important role in this study is played by the time-continuous analogue of iteration, represented by one-parameter semigroups in $\mathsf{Hol}(\mathbb{D}, \mathbb{D})$. For further details and the proofs of the results stated in this subsection we refer the reader to [13, 14] and eventually to references cited there; see also [42, Chapter 3].

Definition 3.5 A family $(\phi_t)_{t\geqslant 0} \subset \mathsf{Hol}(\mathbb{D}, \mathbb{D})$ is said to be a *one-parameter semigroup* if $t \mapsto \phi_t$ is a continuous semigroup homomorphism from the $([0, +\infty), \cdot + \cdot)$ with the usual Euclidian topology to $(\mathsf{Hol}(\mathbb{D}, \mathbb{D}), \cdot \circ \cdot)$ endowed with the topology of locally uniform convergence in $\mathbb{D}$.

Equivalently, (ϕ_t) is a one-parameter semigroup if $\phi_0 = \mathsf{id}_\mathbb{D}$, $\phi_{t+s} = \phi_t \circ \phi_s = \phi_s \circ \phi_t$ for any $t, s \geqslant 0$, and if $\phi_t(z) \to z$ as $t \to 0^+$ for any $z \in \mathbb{D}$.

All elements of a one-parameter semigroup (ϕ_t) different from $\mathsf{id}_\mathbb{D}$ have the same Denjoy–Wolff point and the same set of boundary fixed points. Moreover, if a boundary fixed point σ is regular for *some* of such ϕ_t's, then σ is regular for all ϕ_t's.

Remark 3.6 It is known [5] that for any one-parameter semigroup (ϕ_t) the limit

$$G(z) := \lim_{t \to 0^+} \frac{\phi_t(z) - z}{t}$$

exists for all $z \in \mathbb{D}$ and it is a holomorphic function in $\mathbb{D}$, referred to as the *infinitesimal generator* of (ϕ_t). Moreover, G admits the following representation

$$G(z) = (\tau - z)(1 - \overline{\tau}z)p(z), \quad z \in \mathbb{D}, \tag{3.1}$$

where $\tau \in \overline{\mathbb{D}}$ and $p \in \mathsf{Hol}(\mathbb{D}, \mathbb{C})$ with $\mathsf{Re}\, p \geqslant 0$ are determined by (ϕ_t) uniquely unless $\phi_t = \mathsf{id}_\mathbb{D}$ for all $t \geqslant 0$: namely, τ is the Denjoy–Wolff point of each ϕ_t different from $\mathsf{id}_\mathbb{D}$. Furthermore, for any $z \in \mathbb{D}$, the function $w = w_z(t) := \phi_t(z)$, $t \geqslant 0$, is the unique solution to the initial value problem

$$\frac{dw(t)}{dt} = G(w(t)), \quad t \geqslant 0; \qquad w(0) = z. \tag{3.2}$$

Conversely, if $G \in \mathsf{Hol}(\mathbb{D}, \mathbb{C})$ and for any $z \in \mathbb{D}$ the unique solution to (3.2) extends to all $t \geqslant 0$ (i.e., in other words, G is a holomorphic semicomplete autonomous vector field in $\mathbb{D}$), then G is of the form (3.1) and hence corresponds via (3.2) to a uniquely defined one-parameter semigroup, which we will call *generated by G* and will denote by (ϕ_t^G).

Representation (3.1) is known as the *Berkson–Porta formula*.

3.3   Basics of Modern Loewner Theory

An elementary, but important consequence of Remark 3.6 is that all elements of any one-parameter semigroup in $\mathsf{Hol}(\mathbb{D}, \mathbb{D})$ are *univalent*. However, it is known, see, e.g. [19], that not every univalent $\phi : \mathbb{D} \to \mathbb{D}$ is an element of a one-parameter semigroup. According to Loewner's idea, in order to embed a given univalent self-map ϕ into a semiflow, one should consider a *non-autonomous* version of (3.2), i.e.

$$\frac{dw(t)}{dt} = G\big(w(t), t\big), \quad t \geqslant s \geqslant 0; \qquad w(s) = z \in \mathbb{D}. \tag{3.3}$$

But for which class of functions $G : \mathbb{D} \times [0, +\infty)$ are the corresponding (non-autonomous) semiflows of (3.3) defined globally, i.e. for any $z \in \mathbb{D}$ and any $t \geqslant s \geqslant 0$? Attempts to answer this question have led to the following definition [9, Section 4].

Definition 3.7 A function $G : \mathbb{D} \times [0, +\infty) \to \mathbb{C}$ is called a *Herglotz vector field* (*in the unit disk*) if it satisfies the following conditions:

HVF1.   For every $z \in \mathbb{D}$, the function $[0, +\infty) \ni t \mapsto G(z, t)$ is measurable.

HVF2.   For a.e. $t \in [0, +\infty)$, the function $\mathbb{D} \ni z \mapsto G_t(z) := G(z, t)$ is an infinitesimal generator, i.e. G_t admits the Berkson–Porta representation (3.1).

HVF3.   For any compact set $K \subset \mathbb{D}$ and any $T > 0$ there exists a non-negative locally integrable function $k_{K,T}$ on $[0, +\infty)$ such that $|G(z, t)| \leqslant k_{K,T}(t)$ for all $z \in K$ and a.e. $t \in [0, T]$.

In what follows, we will assume that G in (3.3) is a Herglotz vector field. In such a case by [9, Theorem 4.4], for any $z \in \mathbb{D}$ and any $s \geqslant 0$, the initial value problem (3.3) has a unique solution $w = w_{z,s}^G(t)$ defined for all $t \geqslant s$; and the differential equation in that problem is called the *(generalized) Loewner–Kufarev ODE*. Equation (1.1) in Theorem A and Eq. (3.2) related to one-parameter semigroups are its special cases. (The former equation is often referred as the *(classical radial) Loewner–Kufarev ODE* or simply as the *radial Loewner ODE*.)

Similarly to one-parameter semigroups, the class of all families $(\varphi_{s,t})_{t \geqslant s \geqslant 0}$ which are (non-autonomous) semiflows of Herglotz vector fields G, in the sense that $\varphi_{s,t}(z) = w_{z,s}^G(t)$ for all $z \in \mathbb{D}$, all $s \geqslant 0$, and all $t \geqslant s$, can be characterized intrinsically. Namely, according to [9, Theorem 1.1], initial value problem (3.3) defines a one-to-one correspondence between Herglotz vector fields G and evolution families $(\varphi_{s,t})$, with the latter concept defined as follows.

Definition 3.8 ([9, Definition 3.1]) A family $(\varphi_{s,t})_{t \geqslant s \geqslant 0} \subset \mathsf{Hol}(\mathbb{D}, \mathbb{D})$ is called an *evolution family (in the unit disk)* if it satisfies the following conditions:

EF1. $\varphi_{s,s} = \mathsf{id}_{\mathbb{D}}$ for any $s \geqslant 0$.

EF2. $\varphi_{s,t} = \varphi_{u,t} \circ \varphi_{s,u}$ whenever $0 \leqslant s \leqslant u \leqslant t$.

EF3.   For all $z \in \mathbb{D}$ and $T > 0$ there exists an integrable function $k_{z,T} : [0, T] \to [0, +\infty)$ such that

$$|\varphi_{s,u}(z) - \varphi_{s,t}(z)| \leqslant \int_u^t k_{z,T}(\xi) \, d\xi$$

whenever $0 \leqslant s \leqslant u \leqslant t \leqslant T$.

Now we give a precise definition of what we mean by a Loewner-type parametric representation.

Definition 3.9 ([29, Definition 2.6]) We say that a subsemigroup $\mathfrak{S} \subset \mathsf{Hol}(\mathbb{D}, \mathbb{D})$ *admits Loewner-type parametric representation* if there exists a convex cone $\mathcal{M}_{\mathfrak{S}}$ of Herglotz vector fields in $\mathbb{D}$ with the following properties:

LPR1.   For every $G \in \mathcal{M}_{\mathfrak{S}}$, we have that $G_t := G(\cdot, t)$ for a.e. $t \geqslant 0$ is the generator of a one-parameter semigroup in $\mathfrak{S}$.

LPR2.   The evolution family $(\varphi_{s,t}^G)$ of any $G \in \mathcal{M}_{\mathfrak{S}}$ satisfies $(\varphi_{s,t}^G) \subset \mathfrak{S}$.

LPR3.   For every $\phi \in \mathfrak{S}$ there exists $G \in \mathcal{M}_{\mathfrak{S}}$ such that $\phi = \varphi_{s,t}^G$ for some $s \geqslant 0$ and $t \geqslant s$, where $(\varphi_{s,t}^G)$ stands, as above, for the evolution family of the Herglotz vector field G.

Although the above definition can be stated without using the concept of evolution family, we will take essential advantage of the correspondence between Herglotz vector fields and evolution families. Namely, in the proof of Theorem 1, to check condition LR3 we will use Theorem 2 and results from [29] to construct a suitable evolution family containing ϕ and then prove that its Herglotz vector field belongs to $\mathcal{M}_{\mathfrak{S}}$.

The convex cone $\mathcal{M}_{\mathfrak{S}}$ defined in Remark 2.1 has the property that if $G \in \mathcal{M}_{\mathfrak{S}}$, then $(z, t) \mapsto G(z, at + b)$ also belongs to $\mathcal{M}_{\mathfrak{S}}$ for any $a > 0$ and any $b \geqslant 0$. This fact allows us to replace, in Theorem 1, the condition $\phi = \varphi_{s,t}$ from LPR3 by condition $\phi = \varphi_{0,1}$.

The interplay between evolution families and Herglotz vector fields is completed by constructing related one-parameter families of conformal maps $f_t : \mathbb{D} \to \mathbb{C}$, $t \geqslant 0$, known as Loewner chains.

Definition 3.10 ([16, Definition 1.2]) A family $(f_t)_{t \geqslant 0} \subset \mathsf{Hol}(\mathbb{D}, \mathbb{C})$ is said to be a *Loewner chain (in the unit disk)* if the following three conditions hold:

LC1.   Each function $f_t : \mathbb{D} \to \mathbb{C}$ is univalent.

LC2. $f_s(\mathbb{D}) \subset f_t(\mathbb{D})$ whenever $0 \leqslant s < t$.

LC3. For any compact set $K \subset \mathbb{D}$, there exists a non-negative locally integrable function $k_K : [0, +\infty) \to \mathbb{R}$ such that

$$|f_s(z) - f_t(z)| \leq \int_s^t k_K(\xi) d\xi$$

for all $z \in K$ and all $s, t \geq 0$ with $t \geq s$.

We will use the fact, see [16, Theorem 1.3], that given a Loewner chain (f_t), the functions $\varphi_{s,t} := f_t^{-1} \circ f_s$, $t \geq s \geq 0$, form an evolution family, which is said to be *associated with* the Loewner chain (f_t).

4 Regular Contact Points and Alexandrov–Clark Measure

The following theorem is a slight reformulation of the characterization for existence of finite angular derivative given in [41, pp. 52–53].

Theorem B *Let $\psi \in \mathsf{Hol}(\mathbb{D}, \mathbb{D})$ and let μ be the Alexanderov–Clark measure of ψ at 1, i.e.*

$$\frac{1 + \psi(z)}{1 - \psi(z)} = \int_{\partial \mathbb{D}} \frac{\sigma + z}{\sigma - z} d\mu(\sigma) + iC, \quad \text{for all } z \in \mathbb{D}, \qquad C := \mathsf{Im}\, \frac{1 + \psi(0)}{1 - \psi(0)}.$$

Then for any $\sigma_0 \in \partial \mathbb{D}$, the following two conditions are equivalent:

(i) *σ_0 is a regular contact point of ψ and $\psi(\sigma_0) \neq 1$;*
(ii) *the function $\partial \mathbb{D} \ni \sigma \mapsto |\sigma - \sigma_0|^{-2}$ is μ-integrable.*

Corollary 1 *Let $\Psi \in \mathsf{Hol}(\mathbb{H}, \mathbb{H})$ be given by*

$$\Psi(\zeta) = \alpha + \beta\zeta + \int_{\mathbb{R}} \frac{1 + t\zeta}{t - \zeta} d\nu(t) \quad \text{for all } \zeta \in \mathbb{H}, \qquad (4.1)$$

where $\alpha \in \mathbb{R}$ and $\beta \geq 0$ are some constants and ν is a finite Borel measure on $\mathbb{R}$. If $x_0 \in \mathbb{R}$ is a regular contact point with $\Psi(x_0) \neq \infty$, then $\mathbb{R} \ni t \mapsto |1 + tx_0|/|t - x_0|$ is ν-integrable and

$$\Psi(x_0) = \alpha + \beta x_0 + \int_{\mathbb{R}} \frac{1 + tx_0}{t - x_0} d\nu(t).$$

Proof Write $\psi := H^{-1} \circ \Psi \circ H$, where $H(z) := i(1 + z)/(1 - z)$ is the Cayley transform of $\mathbb{D}$ onto $\mathbb{H}$. Using the relation between the measure ν and the Alexandrov–Clark measure μ of ψ at 1, see, e.g. [6, p. 138], we immediately deduce from Theorem B that $t \mapsto |t - x_0|^{-2}$ is ν-integrable. Since ν is finite, it follows that the functions $t \mapsto |1 + tx_0|/|t - x_0|$, $t \mapsto |t|/|t - x_0|^2$ and $t \mapsto (1 + t^2)/|t - x_0|^2$

are also ν-integrable. It remains to separate the real and imaginary parts in (4.1) for $\zeta := x_0 + i\varepsilon$, $\varepsilon > 0$, and use Lebesgue's Dominated Convergence Theorem to pass to the limit as $\varepsilon \to 0^+$.   □

5 Lemmas

Let us fix a finite set $F \subset \partial\mathbb{D} \setminus \{1\}$ with $n := \mathsf{Card}(F) \geqslant 3$. Write F as a finite sequence $(\sigma_j)_{j=1}^n$ satisfying $0 < \arg\sigma_1 < \arg\sigma_2 < \ldots \arg\sigma_n < 2\pi$. Denote by L_j, $j = 1, \ldots, n-1$, the open arc of $\partial\mathbb{D}$ going from σ_j to σ_{j+1} in the counter-clockwise direction and let L_0 stand for the arc between σ_1 and σ_n containing the point 1.

Let $\phi \in \mathfrak{U}_1[F] \setminus \{\mathsf{id}_\mathbb{D}\}$. Denote by $\mathcal{C}(L)$ the set of all $\psi \in \mathsf{Hol}(\mathbb{D}, \mathbb{D})$ that extend continuously to an open arc $L \subset \partial\mathbb{D}$ with $\psi(L) \subset \partial\mathbb{D}$. By [29, Lemma 5.6(ii)], there is no $j = 1, \ldots, n-1$ such that $\phi \in \mathcal{C}(L_j)$. It may, however, happen that $\phi \in \mathcal{C}(L_0)$.

Remark 5.1 Suppose now that f is any conformal mapping of $\mathbb{D}$ with $\phi(\mathbb{D}) \subset f(\mathbb{D}) \subset \mathbb{D}$. Then for each $j = 1, \ldots, n$ there exists a unique regular contact point ξ_j of f such that $f(\xi_j) = \sigma_j$, see, e.g. [37, Theorem 4.14 on p. 83]. Throughout the paper, we will use this statement implicitly and write $f^{-1}(\sigma_j)$ instead of ξ_j, most of the times without a reference to this remark. Similarly, by $f^{-1}(1)$ we denote the unique regular contact point at which the angular limit of f equals 1.

Remark 5.2 It follows immediately from the above remark and [29, Lemma 5.1] that if $\phi(\mathbb{D}) \subset f_1(\mathbb{D}) \subset f_2(\mathbb{D}) \subset \mathbb{D}$ for some univalent maps $f_1, f_2 : \mathbb{D} \to \mathbb{C}$, then $\psi := f_2^{-1} \circ f_1$ has a regular contact point at $f_1^{-1}(\sigma)$ for any $\sigma \in F \cup \{1\}$ and $\psi\left(f_1^{-1}(\sigma)\right) = f_2^{-1}(\sigma)$.

First, applying [29, Lemma 5.5], we are going to construct a multi-parameter "inclusion chain" $(g_a)_{a \in [0,1]^n}$, which later will be used to obtain an evolution family in $\mathfrak{U}_1[F]$ containing ϕ. In what follows, a and b denote elements of $[0, 1]^n$ with components $a_0, a_1, \ldots, a_{n-1}$ and $b_0, b_1, \ldots, b_{n-1}$, respectively. By $|L|$ we will denote the length of a circular arc or a line segment L.

Lemma 5.3 *In the above notation, there exists a family $(g_a)_{a \in [0,1]^n}$ of univalent holomorphic self-maps of $\mathbb{D}$ satisfying the following conditions:*

(g1) *for every $a \in [0, 1]^n$, $g_a \in \mathfrak{U}[\{1, \sigma_1, \sigma_n\}]$;*

(g2) *$g_{(0,\ldots,0)} = \phi$ and $g_{(1,\ldots,1)} = \mathsf{id}_\mathbb{D}$;*

(g3) *if $a, b \in [0, 1]^n$, $a_j \leqslant b_j$ for all $j = 0, \ldots, n-1$, then $g_a(\mathbb{D}) \subset g_b(\mathbb{D})$, with the strict inclusion if $a_k < b_k$ for some $k \in [1, n-1] \cap \mathbb{N}$;*

(g4) *if $k \in [0, n-1] \cap \mathbb{N}$ and if $a, b \in [0, 1]^n$ with $a_k \leqslant b_k$ and $a_j = b_j$ for all $j = 0, \ldots, n-1$, $j \neq k$, then $\psi_{a,b} := g_b^{-1} \circ g_a \in \mathcal{C}\left(\partial\mathbb{D} \setminus C_k(a)\right)$, with $\psi_{a,b}(\partial\mathbb{D} \setminus C_k(a)) = \partial\mathbb{D} \setminus C_k(b)$, where for $k > 0$ and $x \in [0, 1]^n$, $C_k(x)$ denotes the closed arc of $\partial\mathbb{D}$ going counter-clockwise from $g_x^{-1}(\sigma_k)$ to $g_x^{-1}(\sigma_{k+1})$ and $C_0(x) := L_0$ for any $x \in [0, 1]^n$;*

(g5) g_a *does not depend on* a_0 *if and only if* $\phi \in C(L_0)$; *and if these two equivalent assertions fail, then the inequality* $a_0 < b_0$ *becomes a sufficient condition for the strict inclusion in (g3).*

(g6) *for any map* $t \in [0, +\infty) \mapsto a(t) = (a_0(t), \ldots, a_{n-1}(t)) \in [0, 1]^n$ *with absolutely continuous and non-decreasing components* $t \mapsto a_j(t)$, $j = 0, \ldots, n - 1$, *the functions* $\varphi_{s,t} := g_{a(t)}^{-1} \circ g_{a(s)}$, $t \geq s \geq 0$, *form an evolution family;*

(g7) *the maps* $[0, 1]^n \ni a \mapsto g_a'(1)$ *and* $[0, 1]^n \ni a \mapsto g_a^{-1}(\sigma_j)$, $j = 2, \ldots, n - 1$, *are separately continuous in each of the variables* $a_0, a_1, \ldots, a_n$.

Proof Let us first suppose that ϕ fails to extend continuously to L_0 with $\phi(L_0) \subset \partial \mathbb{D}$. Then by [29, Lemma 5.5], applied with $z_0 := 0$ and $L_n := L_0$, there exists a family $(f_a)_{a \in [0,1]^n} \subset \mathfrak{U}$ such that:

(A) assertions (g3), (g4), and (g5) hold with (g_a) replaced by (f_a);
(B) $f_{(0,\ldots,0)} = \phi$ and $f_{(1,\ldots,1)} \in \mathsf{Aut}(\mathbb{D})$;
(C) $f_a'(0) = \phi(0)$ and $f_a'(0)\overline{\phi'(0)} > 0$ for all $a \in [0, 1]^n$;
(D) the map $[0, 1]^n \ni a \mapsto R(f_a(\mathbb{D}), \phi(0))$, where $R(D, w_0)$ stands for the conformal radius of D w.r.t. w_0, is Lipschitz continuous.

For each $a \in [0, 1]^n$, consider the unique $h_a \in \mathsf{Aut}(\mathbb{D})$ that takes the points $\sigma_0 := 1$, σ_1, and σ_n to the points $f_a^{-1}(1), f_a^{-1}(\sigma_1)$, and $f_a^{-1}(\sigma_n)$, respectively. Then the family (g_a) defined by $g_a := f_a \circ h_a$ satisfy conditions (g1)–(g5).

To prove (g6), we notice that from (A), (C), and (D) it follows that $(F_t)_{t \geq 0}$ defined by $F_t := f_{a(t)}$ for all $t \in [0, +\infty)$ is a Loewner chain, see, e.g. [12, proof of Theorem 2.3]. Let $(\Phi_{s,t})$ be the evolution family associated with (F_t). Then $\eta_j := f_{a(0)}^{-1}(\sigma_j)$, $j = 0, \ldots, n$, are regular contact points of $(\Phi_{s,t})$ in the sense of [11, Definition 3.1]. By [11, Theorem 3.5], the maps $[0, +\infty) \ni t \mapsto \Phi_{0,t}(\eta_j) = f_t^{-1}(\sigma_j)$, $j = 0, \ldots, n$, are locally absolutely continuous. Therefore, with the help of [16, Lemma 2.8] we conclude that the functions

$$\varphi_{s,t} := g_{a(t)}^{-1} \circ g_{a(s)} = h_{a(t)}^{-1} \circ \Phi_{s,t} \circ h_{a(s)}, \quad t \geq s \geq 0,$$

form an evolution family.

It remains to prove (g7). To this end fix some $k \in [0, n-1] \cap \mathbb{N}$ and apply the above argument to the function $t \mapsto a(t) = (a_0(t), \ldots, a_{n-1}(t))$ with the components $a_k = \min\{t, 1\}$ and $a_j = a_j^0$ for all $j = 0, \ldots, n - 1, j \neq k$, where a_j^0's are some arbitrary numbers in $[0, 1]$. As above, by [11, Theorem 3.5], it follows that $t \mapsto |f_{a(t)}'(1)| = |f_{a(0)}'(1)|/|\Phi_{0,t}'(\eta_0)|$ and $t \mapsto f_{a(t)}^{-1}(\sigma_j) = \Phi_{0,t}(\eta_j)$, $j = 0, \ldots, n$, are continuous. This fact immediately implies (g7).

The proof for the case in which ϕ extends continuously to L_0 with $\phi(L_0) \subset \partial \mathbb{D}$ is similar except that the family (g_a) we construct depends only on $n - 1$ parameters a_j, each corresponding to one of the arcs $L_1, \ldots, L_{n-1}$. In this case, for each $a \in [0, 1]^n$, $g_a \in C(L_0)$ with $g_a(L_0) = L_0$. Adding the parameter a_0 formally gives the family (g_a) satisfying all the conditions (g1)–(g7). $\qquad\square$

Now we are going to reduce the number of the parameters by choosing a_0 to be a suitable function of $a' := (a_1, \ldots, a_{n-1})$. As usual, we will identify (a_0, a') with $(a_0, a_1, \ldots, a_{n-1})$.

Lemma 5.4 *Let, as above, $\phi \in \mathfrak{U}_1[F] \setminus \{\mathrm{id}_{\mathbb{D}}\}$ and let $(g_a)_{a \in [0,1]^n}$ be a family of univalent holomorphic self-maps of $\mathbb{D}$ satisfying conditions (g1)–(g5) and (g7) in Lemma 5.3. Then there exists a map $[0, 1]^{n-1} \ni a' := (a_1, \ldots, a_{n-1}) \mapsto a_0(a') \in [0, 1]$ with the following properties:*

(n1) $a_0(0, \ldots, 0) = 0$ and $a_0(1, \ldots, 1) = 1$;
(n2) $a' \mapsto a_0(a')$ *is continuous and non-decreasing in each a_j, $j = 1, \ldots, n-1$;*
(n3) $a' \mapsto \lambda\big(a_0(a'), a'\big)$, *where $\lambda(a) := g_a'(1)$, is non-decreasing in each a_j, $j = 1, \ldots, n-1$.*

Proof Fix for a moment $a, b \in [0, 1]^n$ with $a_j \leq b_j$ for all $j = 0, \ldots, n-1$. Then $\psi_{a,b} := g_b^{-1} \circ g_a \in \mathfrak{U}[1, \sigma_1, \sigma_n]$ thanks to conditions (g1)–(g3) and Remark 5.2. Moreover, by the Chain Rule for angular derivatives, see Remark 3.4, $\lambda(a) = \lambda(b)\psi_{a,b}'(1)$.

Suppose first that $\phi \in C(L_0)$. Then g_a does not depend on a_0 and $\psi_{a,b} \in C(L_0)$. By [29, Lemma 5.6(i)], $\psi_{a,b}'(1) \leq 1$ and hence $\lambda(a) \leq \lambda(b)$. This shows that λ is non-decreasing in variables $a_1, \ldots, a_{n-1}$ and (trivially) does not depend on a_0. Therefore, in case $\phi \in C(L_0)$, we may choose any map $a' \mapsto a_0(a')$ satisfying (n1) and (n2), e.g.,

$$a_0(a') := \left(\textstyle\sum_{j=1}^{n-1} a_j\right) / (n-1).$$

So from now on we may suppose that $\phi \notin C(L_0)$. Let $a_0 = b_0$. Then $\psi_{a,b} \in C(L_0)$ by (g4) and arguing as above, we see that $\psi_{a,b}'(1) \leq 1$. Moreover, if additionally $a \neq b$, then by (g3), $\psi_{a,b} \neq \mathrm{id}_{\mathbb{D}}$ and hence [29, Lemma 5.6(i)] yields the strict inequality $\psi_{a,b}'(1) < 1$. Therefore, λ is increasing in variables $a_1, \ldots, a_{n-1}$. Now let $a' = b' := (b_1, \ldots, b_{n-1})$ and $a_0 < b_0$. Arguing in a similar way, but using the second assertion of [29, Lemma 5.6], we see that $\psi_{a,b}'(1) > 1$, and hence λ is strictly decreasing in a_0. Moreover, recall that by (g7), λ is continuous separately in each variable $a_0, a_1, \ldots, a_{n-1}$. Finally, note that separate continuity and monotonicity in each variable imply joint continuity of $a \mapsto \lambda(a)$, see, e.g. [29, Remark 5.7].

Fix some $a' \in [0, 1]^{n-1}$. Recalling that $\lambda(1, 1, \ldots, 1) = 1$, we have

$$\lambda(1, a') \leq \min\{\lambda(0, a'), 1\} \leq \lambda(0, a').$$

Therefore, because of monotonicity and continuity in a_0, there exists a unique $a_0(a') \in [0, 1]$ such that

$$\lambda(a_0(a'), a') = \min\{\lambda(0, a'), 1\}. \tag{5.1}$$

As the minimum of two continuous non-decreasing functions, the right-hand side of (5.1) is continuous and non-decreasing in $a_1, \ldots, a_{n-1}$, which proves (n3) and continuity of $a' \mapsto a_0(a')$. To see that this map is monotonic, note that by construction $a_0(a') = 0$ if $\lambda(0, a') \leqslant 1$ and that $a_0 = a_0(a') \in [0, 1]$ solves the equation $\lambda(a_0, a') = 1$ if $\lambda(0, a') > 1$. Fix $a', b' \in [0, 1]^{n-1}$ with $a_j \leqslant b_j$ for all $j = 1, \ldots, n-1$. Taking into account that it is not possible to have $\lambda(0, a') > 1$ and $\lambda(0, b') \leqslant 1$ at the same time, careful analysis of the remaining three cases shows that $a_0(a') \leqslant a_0(b')$.

By (g2), $\lambda(0, 0, \ldots, 0) = \phi'(1) \leqslant 1$ and hence $a_0(0, \ldots, 0) = 0$. Similarly, again by (g2), $\lambda(1, 1, \ldots, 1) = 1$. Therefore, $\lambda(0, 1, \ldots, 1) > 1$ and we can easily conclude that $a_0(1, \ldots, 1) = 1$. This completes the proof. $\square$

The following lemma can be viewed as one of the possible analogues of Loewner's Lemma for mappings with boundary Denjoy–Wolff point. Fix some points $\xi_j \in \partial\mathbb{D}, j = 1, \ldots, n$, ordered counter-clockwise in such a way that the arc of $\partial\mathbb{D}$ between ξ_1 and ξ_n that contains $\tau = 1$ does not contain the points $\xi_2, \ldots, \xi_{n-1}$. Denote by $L'_j, j = 1, \ldots, n-1$, the open arc of $\partial\mathbb{D}$ going counter-clockwise from ξ_j to ξ_{j+1}.

Lemma 5.5 *In the above notation, let $\psi \in \mathfrak{U}_1[\{\xi_1, \xi_n\}] \setminus \{\mathrm{id}_\mathbb{D}\}$ and suppose that $\xi_2, \ldots, \xi_{n-1}$ are regular contact points of ψ. Let $\Psi := H \circ \psi \circ H^{-1}$ and $x_j := H(\xi_j)$, $j = 1, \ldots, n$, where $H(z) := i(1 + z)/(1 - z)$ is the Cayley transform of $\mathbb{D}$ onto the upper half-plane $\mathbb{H}$. Let $k \in [1, n-1] \cap \mathbb{N}$. If $\psi \in C(L'_j)$ for all $j = 1, \ldots, n-1$, $j \neq k$, then*

$$\Psi(x_{k+1}) - \Psi(x_k) < x_{k+1} - x_k, \quad \text{and} \tag{5.2}$$

$$\Psi(x_{j+1}) - \Psi(x_j) > x_{j+1} - x_j, \quad \text{for all } j = 1, \ldots, n-1, j \neq k. \tag{5.3}$$

Proof The Nevalinna representation of Ψ is, see, e.g. [6, pp. 135–142]:

$$\Psi(\zeta) = \alpha + \beta\zeta + \int_\mathbb{R} \frac{1 + t\zeta}{t - \zeta} \, d\nu(t) \quad \text{for all } \zeta \in \mathbb{H}, \tag{5.4}$$

where $\alpha \in \mathbb{R}$, $\beta := 1/\phi'(1) \geqslant 1$ and ν is a finite positive Borel measure on $\mathbb{R}$. In accordance with Corollary 1, for $j = 1, \ldots, n-1$ we have

$$\Psi(x_{j+1}) - \Psi(x_j) = (x_{j+1} - x_j)\left[\beta + \int_\mathbb{R} \frac{1 + t^2}{(t - x_j)(t - x_{j+1})} \, d\nu(t)\right]. \tag{5.5}$$

If $j \neq k$, then by hypothesis, $\psi \in C(L'_j)$, and hence $\nu([x_j, x_{j+1}]) = 0$. With this taken into account, (5.5) implies (5.3) with the sign $>$ replaced by $\geqslant$. To check that actually the strict inequality holds, we note that the equality is possible only if $\beta = 1$ and $\nu(\mathbb{R}) = 0$. In such a case, taking into account that $\Psi(x_1) = x_1$, from (5.4) we would get $\Psi = \mathrm{id}_\mathbb{H}$, which contradicts the hypothesis. Finally (5.2) follows immediately because

$$\sum_{j=1}^{n-1} (\Psi(x_{j+1}) - \Psi(x_j)) = \Psi(x_n) - \Psi(x_1) = x_n - x_1 = \sum_{j=1}^{n-1}(x_{j+1} - x_j). \qquad \square$$

Applying the above lemma, we can now pass in the multi-parameter family $(g_a)_{a \in [0,1]}$ to a unique parameter in such a way that the points $\sigma_j, j = 1, \ldots, n$, are kept fixed and the angular derivative at $\tau = 1$ is non-increasing. For $a \in [0,1]^n$ and $j = 1, \ldots, n$, denote $\xi_j(a) := g_a^{-1}(\sigma_j)$.

Lemma 5.6 *Under hypothesis of Lemma 5.4 there exists a continuous map $[0,1] \ni \theta \mapsto a(\theta) = \big(a_0(\theta), a_1(\theta), \ldots, a_n(\theta)\big) \in [0,1]^n$ such that:*

(m1) $\theta \mapsto a_j(\theta)$ *is strictly increasing for each $j = 1, \ldots n$ and non-decreasing for* $j = 0$;
(m2) $a(0) = (0, \ldots, 0)$ *and* $a(1) = (1, \ldots, 1)$;
(m3) $\xi_j\big(a(\theta)\big) = \sigma_j$ *for each $j = 1 \ldots, n$ and any $\theta \in [0,1]$;*
(m4) $\theta \mapsto \lambda(a(\theta)) = g'_{a(\theta)}(1)$ *is non-decreasing.*

Proof Consider the family $(h_{a'})_{a' \in [0,1]^{n-1}}$, defined by $h_{a'} := g_{(a_0(a'),a')}$, where $a' \mapsto a_0(a')$ is the map defined in Lemma 5.4.

Given $a', b' \in [0,1]^{n-1}$ with $a_j \le b_j$ for $j = 1, \ldots, n$, consider the map $\psi_{a',b'} := h_{b'}^{-1} \circ h_{a'}$. The points $\tau = 1$ and $\xi_j(a') := h_{a'}^{-1}(\sigma_j), j = 1, \ldots, n$, are regular contact points of $\psi_{a',b'}$, with $\psi_{a',b'}(1) = 1$ and $\psi_{a',b'}(\xi_j(a')) = \xi_j(b')$, see Remark 5.2. In particular,

$$\xi_1(a') = \xi_1(b') = \sigma_1, \qquad \xi_n(a') = \xi_n(b') = \sigma_n$$

and hence $\psi_{a',b'} \in \mathfrak{U}[1, \sigma_1, \sigma_n]$. Taking into account that

$$h'_{a'}(1) = \lambda(a_0(a'), a') \le \lambda(a_0(b'), b') = h'_{b'}(1)$$

by Lemma 5.4, we conclude that $\psi_{a',b'} \in \mathfrak{U}_1[\sigma_1, \sigma_n]$. Moreover, thanks to condition (g3), $\psi_{a',b'} \ne \mathrm{id}_{\mathbb{D}}$ unless $a' = b'$.

Applying Lemma 5.5 to the map $\psi := \psi_{a',b'}$ with regular contact points $\xi_j := \xi_j(a')$, we see that for any $k = 1, \ldots, n-1$, the function

$$a' = (a_1, \ldots, a_{n-1}) \mapsto \ell_k(a') := H(\xi_{k+1}(a')) - H(\xi_k(a'))$$

is strictly decreasing in a_k and strictly increasing in each a_j with $j \ne k$. Moreover, by (g7), this function is separately continuous in each variable, which in view of monotonicity, implies the joint continuity, see, e.g. [29, Remark 5.7].

To prove the lemma, it is sufficient to construct a continuous map

$$[0,1] \ni \theta \mapsto a'(\theta) = \big(a_1(\theta), \ldots, a_{n-1}(\theta)\big) \in [0,1]^{n-1}$$

satisfying the following conditions:

(p1) all the components $\theta \mapsto a_j(\theta), j = 1, \ldots, n-1$, are strictly increasing;
(p2) $a'(0) = (0, \ldots, 0), a'(1) = (1, \ldots, 1)$, and
(p3) the maps $\theta \mapsto \ell_j(a'(\theta)), j = 1, \ldots, n-1$, are constant.

The construction is based on a recursive procedure. First we will find a continuous function $[0, 1]^{n-2} \ni a'' := (a_1, \ldots, a_{n-2}) \mapsto a_{n-1}(a'') \in [0, 1]$, $a_{n-1}(0, \ldots, 0) = 0$, $a_{n-1}(1, \ldots, 1) = 1$, strictly increasing in each variable and such that $\ell_{n-1}(a'', a_{n-1}(a''))$ is constant. Namely, for any $a'' \in [0, 1]^{n-2}$ we set $a_{n-1}(a'')$ to be a solution to

$$\ell_{n-1}(a'', a_{n-1}) = H(\sigma_n) - H(\sigma_{n-1}),$$

which exists and unique thanks to the continuity and monotonicity of ℓ_{n-1} in a_{n-1} and to the fact that by monotonicity in $a_1, \ldots, a_{n-2}$ we have

$$\ell_{n-1}(a'', 0) \geqslant \ell_{n-1}(0, \ldots, 0) = H(\sigma_n) - H(\sigma_{n-1}) \quad \text{and} \tag{5.6}$$

$$\ell_{n-1}(a'', 1) \leqslant \ell_{n-1}(1, \ldots, 1) = H(\sigma_n) - H(\sigma_{n-1}). \tag{5.7}$$

The required properties of continuity and monotonicity of $a'' \mapsto a_{n-1}(a'')$ follow from continuity and monotonicity of ℓ_{n-1}. Finally, using again (5.6) and (5.7) one immediately obtains equalities $a_{n-1}(0, \ldots, 0) = 0$ and $a_{n-1}(1, \ldots, 1) = 1$.

For $n = 3$ the proof can be now completed by taking

$$a(\theta) := \big(a_0(\theta, a_2(\theta)), \theta, a_2(\theta)\big)$$

for all $\theta \in [0, 1]$. So suppose that $n > 3$. It is easy to see that for any $k = 1, \ldots, n-2$, the function $\ell_k^1(a'') := \ell_k(a'', a_{n-1}(a''))$, defined for all $a'' := (a_1, \ldots, a_{n-2}) \in [0, 1]^{n-2}$, is continuous and strictly increasing in each a_j with $j \neq k$. Since

$$\sum_{j=1}^{n-2} \ell_j^1(a'') = H(\sigma_n) - H(\sigma_1) - \ell_{n-1}\big(a'', a_{n-1}(a'')\big) = H(\sigma_{n-1}) - H(\sigma_1),$$

it follows immediately that ℓ_k^1 is strictly decreasing in a_k. Furthermore,

$$\ell_{n-2}^1(0, \ldots, 0) = \ell_{n-2}(0, \ldots, 0, 0) = H(\sigma_{n-1}) - H(\sigma_{n-2}) \quad \text{and}$$

$$\ell_{n-2}^1(1, \ldots, 1) = \ell_{n-2}(1, \ldots, 1, 1) = H(\sigma_{n-1}) - H(\sigma_{n-2}).$$

Therefore, the above argument can be applied again with ℓ_{n-1} replaced by ℓ_{n-2}^1. In other words, we exclude one more parameter by finding $a_{n-2} = a_{n-2}(a''') \in [0, 1]$, $a''' := (a_1, \ldots, a_{n-3})$, that solves the equation $\ell_{n-2}^1(a''', a_{n-2}) = H(\sigma_{n-1}) - H(\sigma_{n-2})$.

Repeating this procedure suitable number of times, we end up with a continuous map $[0, 1] \ni \theta \mapsto a'(\theta) = \big(\theta, a_2(\theta), \ldots, a_{n-1}(\theta)\big) \in [0, 1]^{n-1}$ that satisfies (p1)–(p3). Using conditions (n1)–(n3) in Lemma 5.4, we see that the map $\theta \mapsto a(\theta) := \big(a_0(a'(\theta)), a'(\theta)\big)$ satisfies (m1)–(m4). The proof is now complete. $\qquad \square$

6 Proof of Main Results

6.1 Proof of Theorem 2

Without loss of generality we may assume that $\tau = 1$. Furthermore, for $\phi = \mathrm{id}_{\mathbb{D}}$ the statement of the theorem is trivial, so will suppose that $\phi \neq \mathrm{id}_{\mathbb{D}}$.

Now we can apply Lemma 5.3 to construct the multi-parameter family $(g_a)_{a \in [0,1]^n}$. Next using Lemma 5.6, we obtain the continuous map $[0,1] \ni \theta \mapsto a = \big(a_0(\theta), \ldots, a_{n-1}(\theta)\big)$. By (m1) and (m2), the function

$$\Lambda(\theta) := \frac{1}{n} \sum_{j=0}^{n-1} a_j(\theta), \qquad \theta \in [0,1],$$

is a continuous strictly increasing map of $[0,1]$ onto itself. Let $[0,1] \ni t \mapsto \theta(t)$ be the inverse of the function Λ, which we extend to $[0, +\infty)$ by setting $\theta(t) := 1$ for all $t > 1$. Clearly, the functions $t \mapsto a_j(\theta(t))$, $j = 0, \ldots, n-1$, are non-decreasing and

$$|a_j(\theta(t_2)) - a_j(\theta(t_1))| \leqslant \sum_{k=0}^{n-1} |a_k(\theta(t_2)) - a_k(\theta(t_1))| \leqslant |t_2 - t_1|$$

for any $t_1, t_2 \geqslant 0$ and any $j = 0, \ldots, n-1$. Therefore, by assertion (g6) of Lemma 5.3, the functions

$$\varphi_{s,t} := g_{a(\theta(t))}^{-1} \circ g_{a(\theta(t))}, \qquad t \geqslant s \geqslant 0,$$

form an evolution family. Moreover, since $a(\theta(0)) = (0, \ldots, 0)$ and $a(\theta(1)) = (1, \ldots, 1)$, by assertion (g2) of Lemma 5.3 we have $\varphi_{0,1} = \phi$.

Bearing in mind Remarks 3.4 and 5.2, it remains to mention that $(\varphi_{s,t}) \subset \mathfrak{U}[F]$ by (m3) and that $(\varphi_{s,t}) \subset \mathfrak{U}_1$ thanks to (m4). $\square$

6.2 Proof of Theorem 1

It is sufficient to prove the theorem for $\mathfrak{S} = \mathfrak{U}[F]$ or for $\mathfrak{S} = \mathfrak{U}_\tau[F]$, where $F \subset \partial \mathbb{D}$ is a finite set with $\mathsf{Card}(F) \geqslant 3$ and $\tau \in \partial \mathbb{D} \setminus F$, as all other cases are already covered by [29, Theorem 1].

Fix any Herglotz vector field G. By the very definition, for a.e. $s \geqslant 0$, $G(\cdot, s)$ is an infinitesimal generator. By [15, Theorem 1], the one-parameter semigroup $(\phi_t^{G(\cdot,s)})$ generated by $G(\cdot, s)$ has a boundary regular fixed point at $\sigma \in \partial \mathbb{D}$ if and only if there exists a finite angular limit

$$\lambda(\sigma, s) := \angle \lim_{z \to \sigma} \frac{G(z, s)}{z - \sigma}. \qquad (6.1)$$

Now let $\mathcal{M}_{\mathfrak{S}}$ be the set of all Herglotz vector fields G satisfying the following conditions:

(a) for a.e. $s \geqslant 0$, we have $(\phi_t^{G(\cdot, s)}) \subset \mathfrak{S}$;
(b) for every $\sigma \in F$, the function $\lambda(\sigma, \cdot)$ is locally integrable on $[0, +\infty)$.

First of all, let us show that $\mathcal{M}_{\mathfrak{S}}$ is a convex cone. Let $G_1, G_2 \in \mathcal{M}_{\mathfrak{S}} \setminus \{0\}$ and let $G_3 \not\equiv 0$ be a linear combination of G_1 and G_2 with non-negative coefficients. Then using the fact that infinitesimal generators form a convex cone, see, e.g. [1, Corollary 1.4.15 on p. 108], we see that G_3 is a Herglotz vector field. Moreover, if for some $s \geqslant 0$, τ is the Denjoy–Wolff point of both $(\phi_t^{G_1(\cdot, s)})$ and $(\phi_t^{G_2(\cdot, s)})$, then thanks to the Berkson–Porta formula, see Remark 3.6, τ is also the Denjoy–Wolff point of $(\phi_t^{G_3(\cdot, s)})$. Finally, since the angular limit (6.1) exists finitely and satisfies (b) both for $G := G_1$ and for $G := G_2$, this is the case for $G := G_3$ as well. Therefore, $G_3 \in \mathcal{M}_{\mathfrak{S}}$.

It remains to show that $\mathcal{M}_{\mathfrak{S}}$ meets conditions (i)–(iii) in Theorem 1. Firstly, condition (i) simply coincides with (a). Furthermore, in view of [15, Theorem 1] and [11, Theorem 1.1], conditions (a) and (b) imply that the evolution family $(\varphi_{s,t}^G)$ generated by G satisfies $(\varphi_{s,t}^G) \subset \mathfrak{U}[F]$. This proves (iii) in case $\mathfrak{S} = \mathfrak{U}[F]$. Notice that if $\mathfrak{S} = \mathfrak{U}_\tau[F]$, then in view of the Berkson–Porta formula, see Remark 3.6, from (a) it follows that $G(z, t) = (\tau - z)(1 - \overline{\tau}z)p_t(z)$ for all $z \in \mathbb{D}$ and a.e. $t \geqslant 0$, where $(p_t)_{t \geqslant 0}$ is a family of holomorphic functions with non-negative real part. By [9, Corollary 7.2], the latter equality implies that $(\varphi_{s,t}^G) \subset \mathfrak{U}_\tau$. Hence (iii) also holds for the case $\mathfrak{S} = \mathfrak{U}_\tau[F]$.

It remains to show that $\mathcal{M}_{\mathfrak{S}}$ satisfies (ii). Let $\phi \in \mathfrak{S} \setminus \{\mathrm{id}_{\mathbb{D}}\}$. We have to find $G \in \mathcal{M}_{\mathfrak{S}}$ whose evolution family $(\varphi_{s,t}) = (\varphi_{s,t}^G)$ satisfies $\varphi_{0,1} = \phi$. To this end we first construct a certain evolution family $(\varphi_{s,t})$ and then show that its Herglotz vector field is a suitable candidate for G. Let ξ be the Denjoy–Wolff point of ϕ. Of course, $\xi = \tau$ if $\mathfrak{S} = \mathfrak{U}_\tau[F]$.

Claim there exists an evolution family $(\varphi_{s,t}) \subset \mathfrak{U}[F] \cap \mathfrak{U}_\xi$ such that $\varphi_{0,1} = \phi$.

Indeed, if $\xi \in \partial \mathbb{D} \setminus F$, then the Claim follows readily from Theorem 2 applied with τ replaced with ξ. Similarly, if $\xi \in F$ and $\mathrm{Card}(F) > 3$, then we should apply Theorem 2 with τ and F replaced with ξ and $F \setminus \{\xi\}$, respectively. In the remaining cases, i.e. if $\xi \in F$ and $\mathrm{Card}(F) = 3$ or if $\xi \in \mathbb{D}$, the proof of the Claim is contained in the proof of [29, Theorem 1], again with ξ and $F \setminus \{\xi\}$ substituted for τ and F, respectively.

Now by the above Claim, there exists an evolution family $(\varphi_{s,t}) \subset \mathfrak{S}$ with $\varphi_{0,1} = \phi$. By [15, Theorem 1] and [11, Theorem 1.1], the Herglotz vector field G of $(\varphi_{s,t})$

belongs to $\mathcal{M}_{\mathfrak{U}[F]}$. This completes the proof for $\mathfrak{S} = \mathfrak{U}[F]$, while in case $\mathfrak{S} = \mathfrak{U}_\tau[F]$ it remains to notice that $(\phi_t^{G(\cdot,s)}) \subset \mathfrak{U}_\tau$ for a.e. $s \geqslant 0$ by [9, Theorem 6.7]. $\square$

Acknowledgements The author was partially supported by *Ministerio de Economía y Competitividad* (Spain) project MTM2015-63699-P.

References

1. Abate, M.: Iteration theory of holomorphic maps on taut manifolds. Research and Lecture Notes in Mathematics. Complex Analysis and Geometry. Mediterranean, Rende (1989)
2. Aleksandrov, I.A.: Parametric continuations in the theory of univalent functions (Russian). Izdat. Nauka, Moscow (1976). MR0480952
3. Anderson, J.M., Vasil'ev, A.: Lower Schwarz-Pick estimates and angular derivatives. Ann. Acad. Sci. Fenn. Math. **33**(1), 101–110 (2008). MR2386840
4. Arosio, L., Bracci, F., Wold, E.F.: Solving the Loewner PDE in complete hyperbolic starlike domains of $\mathbb{C}^N$. Adv. Math. **242**, 209–216 (2013). MR3055993
5. Berkson, E., Porta, H.: Semigroups of holomorphic functions and composition operators. Michigan Math. J. **25**, 101–115 (1978)
6. Bhatia, R.: Matrix Analysis. Graduate Texts in Mathematics, vol. 169. Springer, New York (1997). MR1477662
7. Bracci, F., Contreras, M.D., Díaz-Madrigal, S.: Aleksandrov-Clark measures and semigroups of analytic functions in the unit disc. Ann. Acad. Sci. Fenn. Math. **33**(1), 231–240 (2008). MR2386848
8. Bracci, F., Contreras, M.D., Díaz-Madrigal, S.: Evolution families and the Loewner equation II: complex hyperbolic manifolds. Math. Ann. **344**(4), 947–962 (2009)
9. Bracci, F., Contreras, M.D., Díaz-Madrigal, S.: Evolution families and the Loewner equation I: the unit disc. J. Reine Angew. Math. (Crelle's J.) **672**, 1–37 (2012)
10. Bracci, F., Contreras, M.D., Díaz-Madrigal, S., Vasil'ev, A.: Classical and stochastic Löwner-Kufarev equations. In: Harmonic and Complex Analysis and Its Applications, pp. 39–134. Trends in Mathematics, Birkhäuser, Cham (2014). MR3203100
11. Bracci, F., Contreras, M.D., Díaz-Madrigal, S., Gumenyuk, P.: Boundary regular fixed points in Loewner theory. Ann. Mat. Pura Appl. (4) **194**(1), 221–245 (2015)
12. Contreras, M.D.: Geometry behind chordal Loewner chains. Complex Anal. Oper. Theory **4**(3), 541–587 (2010). MR2719792
13. Contreras, M.D., Díaz-Madrigal, S.: Analytic flows in the unit disk: angular derivatives and boundary fixed points. Pac. J. Math. **222**, 253–286 (2005)
14. Contreras, M.D., Díaz-Madrigal, S., Pommerenke, C.: Fixed points and boundary behaviour of the Koenigs function. Ann. Acad. Sci. Fenn. Math. **29**(2), 471–488 (2004). MR2097244
15. Contreras, M.D., Díaz-Madrigal, S., Pommerenke, C.: On boundary critical points for semigroups of analytic functions. Math. Scand. **98**(1), 125–142 (2006). MR2221548
16. Contreras, M.D., Díaz-Madrigal, S., Gumenyuk, P.: Loewner chains in the unit disk. Rev. Mat. Iberoam. **26**, 975–1012 (2010)
17. Cowen, C.C., Pommerenke, C.: Inequalities for the angular derivative of an analytic function in the unit disk. J. Lond. Math. Soc. (2) **26**(2), 271–289 (1982). MR0675170
18. de Branges, L.: A proof of the Bieberbach conjecture. Acta Math. **154**(1-2), 137–152 (1985). MR0772434
19. Elin, M., Goryainov, V., Reich, S., Shoikhet, D.: Fractional iteration and functional equations for functions analytic in the unit disk. Comput. Methods Funct. Theory **2**(2), 353–366 (2002), [On table of contents: 2004]. MR2038126

20. Frolova, A., Levenshtein, M., Shoikhet, D., Vasil'ev, A.: Boundary distortion estimates for holomorphic maps. Complex Anal. Oper. Theory **8**(5), 1129–1149 (2014). MR3208806
21. Goryainov, V.V.: Semigroups of conformal mappings. Mat. Sb. (N.S.) **129(171)**(4), 451–472, (1986) (Russian); Translation in Math. USSR Sbornik **57**, 463–483 (1987)
22. Goryainov, V.V.: Fractional iterates of functions that are analytic in the unit disk with given fixed points. Mat. Sb. **182**(9), 1281–1299 (1991); Translation in Math. USSR-Sb. **74**(1), 29–46 (1993). MR1133569
23. Goryainov, V.V., Evolution families of analytic functions and time-inhomogeneous Markov branching processes. Dokl. Akad. Nauk **347**(6), 729–731 (1996); Translation in Dokl. Math. **53**(2), 256–258 (1996)
24. Goryainov, V.V.: Semigroups of analytic functions in analysis and applications. Uspekhi Mat. Nauk **676**(408), 5–52 (2012); Translation in Russian Math. Surveys **67**(6), 975–1021 (2012)
25. Goryainov, V.V.: Evolution families of conformal mappings with fixed points. (Russian. English summary). Zírnik Prats' Instytutu Matematyky NAN Ukrainy. National Academy of Sciences of Ukraine (ISSN 1815-2910) **10**(4-5), 424–431 (2013). Zbl 1289.30024
26. Goryainov, V.V.: Evolution families of conformal mappings with fixed points and the Löwner-Kufarev equation. Mat. Sb. **206**(1), 39–68 (2015); Translation in Sb. Math. **206**(1-2), 33–60 (2015)
27. Goryainov, V.V., Ba, I.: Semigroups of conformal mappings of the upper half-plane into itself with hydrodynamic normalization at infinity. Ukr. Math. J. **44**, 1209–1217 (1992)
28. Goryainov, V.V., Kudryavtseva, O.S.: One-parameter semigroups of analytic functions, fixed points and the Koenigs function. Mat. Sb. **202**(7), 43–74 (2011) (Russian); Translation in Sbornik: Mathematics **202**(7-8), 971–1000 (2011)
29. Gumenyuk, P.: Parametric representation of univalent functions with boundary regular fixed points. Constr. Approx. **46**, 435–458 (2017). https://doi.org/10.1007/s00365-017-9376-4
30. Koch, J., Schleißinger, S.: Value ranges of univalent self-mappings of the unit disc. J. Math. Anal. Appl. **433**(2), 1772–1789 (2016). MR3398791
31. Löwner, K.: Untersuchungen über schlichte konforme Abbildungen des Einheitskreises. Math. Ann. **89**, 103–121 (1923)
32. Loewner, C.: On totally positive matrices. Math. Z. **63**, 338–340 (1955). MR0073657
33. Loewner, C.: Seminars on Analytic Functions, vol. 1. Institute for Advanced Study, Princeton, NJ (1957). Available at http://babel.hathitrust.org
34. Loewner, C.: On generation of monotonic transformations of higher order by infinitesimal transformations. J. Anal. Math. **11**, 189–206 (1963). MR0214711
35. Poggi-Corradini, P.: Canonical conjugations at fixed points other than the Denjoy-Wolff point. Ann. Acad. Sci. Fenn. Math. **25**(2), 487–499 (2000). MR1762433
36. Pommerenke, Ch.: Univalent functions. With a chapter on quadratic differentials by Gerd Jensen. Vandenhoeck & Ruprecht, Göttingen (1975)
37. Pommerenke, Ch.: Boundary Behaviour of Conformal Mappings. Springer, New York (1992)
38. Pommerenke, C., Vasil'ev, A.: Angular derivatives of bounded univalent functions and extremal partitions of the unit disk. Pac. J. Math. **206**(2), 425–450 (2002). MR1926785 (2003i:30024)
39. Prokhorov, D., Samsonova, K.: Value range of solutions to the chordal Loewner equation. J. Math. Anal. Appl. **428**(2) (2015). MR3334955
40. Roth, O., Schleißinger, S.: Rogosinski's lemma for univalent functions, hyperbolic Archimedean spirals and the Loewner equation. Bull. Lond. Math. Soc. **46**(5), 1099–1109 (2014). MR3262210
41. Sarason, D.: Sub-Hardy Hilbert Spaces in the Unit Disk. University of Arkansas Lecture Notes in the Mathematical Sciences, vol. 10. Wiley, New York (1994). MR1289670
42. Shoikhet, D.: Semigroups in Geometrical Function Theory. Kluwer, Dordrecht (2001)

Is There a Teichmüller Principle in Higher Dimensions?

Oliver Roth

Abstract The underlying theme of Teichmüller's papers in function theory is a general principle which asserts that every extremal problem for univalent functions of one complex variable is connected with an associated quadratic differential. The purpose of this paper is to indicate a possible way of extending Teichmüller's principle to several complex variables. This approach is based on the Loewner differential equation.

Keywords Loewner theory · Extremal problems · Control theory

Mathematics Subject Classification (2000) 30C55, 32H02, 49K15

1 Introduction

We denote by $\mathrm{Hol}(\mathbb{B}^n, \mathbb{C}^n)$ the set of all holomorphic maps from the open unit ball $\mathbb{B}^n := \{z \in \mathbb{C}^n \,:\, ||z|| < 1\}$ equipped with the standard Euclidean norm $|| \cdot ||$ of $\mathbb{C}^n$ into $\mathbb{C}^n$. Endowed with the compact-open topology of locally uniform convergence, the vector space $\mathrm{Hol}(\mathbb{B}^n, \mathbb{C}^n)$ becomes a Fréchet space. In geometric function theory, the *univalent* maps in $\mathrm{Hol}(\mathbb{B}^n, \mathbb{C}^n)$ are of particular interest and extremal problems provide an effective method for establishing the existence of univalent maps with certain natural properties.

This point of view is particularly successful in the classical one-dimensional case, since the class

$$\mathcal{S} := \{f \in \mathrm{Hol}(\mathbb{D}, \mathbb{C}) \,:\, f(0) = 0, f'(0) = 1, f \text{ univalent}\}$$

O. Roth (✉)
Department of Mathematics, University of Würzburg, Würzburg, Germany
e-mail: roth@mathematik.uni-wuerzburg.de

© Springer International Publishing AG, part of Springer Nature 2017   87
F. Bracci (ed.), *Geometric Function Theory in Higher Dimension*,
Springer INdAM Series 26, https://doi.org/10.1007/978-3-319-73126-1_7

of all normalized univalent (or schlicht) functions on the unit disk $\mathbb{D} := \mathbb{B}^1$ is a compact subset of $\mathrm{Hol}(\mathbb{D}, \mathbb{C})$ and so any continuous functional $J : S \to \mathbb{R}$ attains its maximum value within the class S. This means that there exists at least one $F \in S$ such that $J(f) \leq J(F)$ for any $f \in S$. We call such a map F an *extremal function for J over S*.

Around 1938, Teichmüller [41] stated a general principle which roughly says that any extremal problem over S is associated in a well-defined way with a quadratic differential. Teichmüller did not go on to give a precise formulation of his principle in upmost generality, but he did content himself with applying his principle to a number of specific, yet characteristic special cases. Only much later, Jenkins (see [20]) succeeded in formulating what is called the General Coefficient Theorem and which can be regarded as a rigorous version of Teichmüller's principle for a fairly large class of extremal problems.

In order to state Teichmüller's principle, the following notion is useful.

Definition 1.1 ([25, 40]) Let Q be a meromorphic function on $\mathbb{C}$. A formal expression of the form

$$Q(w) \, dw^2$$

is called a *quadratic differential*. A function $F \in S$ is called *admissible for the quadratic differential* $Q(w) \, dw^2$, if F maps $\mathbb{D}$ onto $\mathbb{C}$ minus a set of finitely many analytic arcs $w = w(t)$ satisfying $Q(w)dw^2 > 0$.

Now, roughly speaking Teichmüller's principle asserts that to each quadratic differential one can associate an extremal problem for univalent functions in such a way that the extremal functions are admissible for this quadratic differential. This principle has a partial converse, first established by Schiffer [36], which says that for any extremal problem for univalent functions one can associate a quadratic differential $Q(w) \, dw^2$ so that the corresponding extremal functions are admissible for $Q(w) \, dw^2$.

For the sake of simplicity, we restrict our discussion to a particular simple, yet important case and consider for a fixed integer $N \geq 2$ the N-th coefficient functional

$$J_N(f) := a_N, \qquad f(z) = z + \sum_{k=2}^{\infty} a_k z^k \in \mathrm{Hol}(\mathbb{D}, \mathbb{C}).$$

For this functional, one can state the Teichmüller–Schiffer paradigma in a precise and simple way, see [8, Chapter 10.8]. We start with the Schiffer differential equation.

Theorem 1.2 (Schiffer's Theorem) Let $F(z) = z + \sum_{k=2}^{\infty} A_k z^k \in S$ be an extremal function for $\mathrm{Re}\, J_N$ over S and let

$$P_N(w) := \sum_{k=1}^{N-1} J_N(F^{k+1}) w^k. \tag{1.1}$$

Then F is admissible for the quadratic differential

$$-P_N\left(\frac{1}{w}\right)\frac{dw^2}{w^2},$$

and F is a solution to the differential equation

$$\left[\frac{zF'(z)}{F(z)}\right]^2 P_N\left(\frac{1}{F(z)}\right) = R_N(z), \tag{1.2}$$

with

$$R_N(z) = (N-1)A_n + \sum_{k=1}^{N-1}\left(kA_k z^{k-N} + k\overline{A_k}z^{N-k}\right) \tag{1.3}$$

and

$$R_N(\kappa) \geq 0 \quad \text{for all } \kappa \in \partial\mathbb{D} \text{ with equality for at least one } \kappa_0 \in \partial\mathbb{D}. \tag{1.4}$$

Remarks 1.3 A few of remarks are in order.

(a) The only functions $F \in \mathcal{S}$ for which Theorem 1.2 actually applies are the Koebe functions $k(z) = z/(1 - \alpha z)^2$ with $\alpha^{N-1} = 1$. This is a consequence of de Branges' theorem [7], that is, the former Bieberbach conjecture, which states that $|a_N| \leq N$ with equality if and only if $F(z) = z/(1 - \eta z)^2$ with $|\eta| = 1$. However, Schiffer's method works for much more general ("differentiable") functionals $J : \mathcal{S} \to \mathbb{C}$ as well, see [8]. In this expository paper, we nevertheless focus mainly on the simple functional J_N for various reasons. First of all, the more general cases of Schiffer's theorem are essentially as difficult to prove as the case J_N. Second, the relation to Teichmüller's principle is most easily described for the functional J_N. Finally, the main issue of this note are extensions to higher dimensions with a view toward a higher dimensional Bieberbach conjecture (see e.g. [3]).

(b) In order to prove Theorem 1.2 one compares the extremal function with nearby functions in the class S. The construction of suitable comparison functions is a nontrivial task, because the family $\mathcal{S}$ is highly nonlinear. In one dimension there are several variational methods for univalent functions available. We mention the work of Schiffer, Schaeffer and Spencer, Goluzin and others, see [8, 25]. These methods are geometric in nature and make use of the Riemann mapping theorem and are thus specifically one dimensional. A different approach is possible by way of the Loewner equation and Pontyagin's Maximum Principle from optimal control. We explain this in more detail below.

(c) Equation (1.2) is called the *Schiffer differential equation*. It is analogous to the Euler equation in the classical calculus of variations. Like the Euler equation, it expresses the fact that every extremal function is a critical point of J_N. However, an additional difficulty arises, because the Schiffer differential equation involves the initial coefficients $A_2, \ldots, A_{N-1}$ of the unknown extremal function.

(d) It is not difficult to show that $F \in S$ satisfies Schiffer's equation (1.2) such that the "positivity condition" (1.4) holds if and only if F is admissible for the quadratic differential $-P_N(1/w)\frac{dw^2}{w^2}$. See the proof of Theorem 1.2 in [8] for the "only if"-part and e.g. [25, Proof of Theorem 7.5] for the "if"-part.

(e) Theorem 1.2 can further be strengthened by showing that if $F \in S$ is extremal for the functional J_N over S, then F is a one-slit map, that is, F maps $\mathbb{D}$ onto $\mathbb{C}$ minus a single analytic arc. Moreover, this arc has several additional geometric properties such as increasing modulus, the $\pi/4$-property and an asymptotic direction at infinity, see [8, Chapter 10] for more on this.

Theorem 1.4 (Teichmüller's Coefficient Theorem [41]) *Let* $P(w) = w^{N-1} + c_{N-2}w^{N-2} + \cdots + c_1 w + c_0$ *be a polynomial. Suppose that*

$$F(z) = z + \sum_{k=2}^{\infty} A_k z^k \in S$$

is admissible for the quadratic differential

$$-P\left(\frac{1}{w}\right)\frac{dw^2}{w^2}.$$

Then

$$\operatorname{Re} J_N(f) \leq \operatorname{Re} J_N(F)$$

for any

$$f \in S(A_2, \ldots, A_{N-1}) := \left\{ f(z) = z + \sum_{k=2}^{\infty} a_k z^k \in S : a_2 = A_2, \ldots, a_{N-1} = A_{N-1} \right\}.$$

Equality occurs only for $f = F$.

In short, under the assumptions of Theorem 1.4, that is, if $F \in S$ is a solution to the Schiffer differential equation (1.2) such that the positivity condition (1.4) holds, then F is an extremal function for the real part of the Bieberbach functional $J_N(f)$, but subject to the *side conditions* $a_2 = A_2, \ldots, a_{N-1} = A_{N-1}$:

Conclusion 1.5 Let $F(z) = z + \sum_{k=2}^{\infty} A_k z^k \in S$. Then the condition that

F is a solution to Schiffer's differential equation (1.2) such that (1.4) holds is

(a) necessary for F being extremal for $\operatorname{Re} J_N$ over the entire class $\mathcal{S}$, and
(b) sufficient for F being extremal for $\operatorname{Re} J_N$ over the restricted class $\mathcal{S}(A_2, \ldots, A_{N-1})$.

Remark 1.6 Teichmüller was quite confident about his result and he conjectured[1] that it can be used to solve the

---

General Coefficient Problem for Univalent Functions

Given $F(z) = z + \sum\limits_{k=2}^{\infty} A_k z^k \in \mathcal{S}$. Find for each $N \in \mathbb{N}$,

$$\left\{ a_N : f(z) = z + \sum_{k=2}^{\infty} a_k z^k \in \mathcal{S}(A_2, \ldots, A_{N-1}) \right\}.$$

The goal of this note is to show that Schiffer's theorem can be extended, at least in spirit, to higher dimensions using the Loewner equation. We also give a statement of Teichmüller's Coefficient Theorem entirely in terms of the Loewner equation. This principally opens up the possibility for an extension of Teichmüller's principle to higher dimensions.

The literature on univalent functions in general, and the Loewner equation in particular, is extensive and there are several excellent survey papers available, see, e.g. [4]. We therefore have included only few references about the subject. As this paper is expository, it contains virtually no proof. An exception is Theorem 6.1.

2 The Loewner Differential Equation and the Class $\mathcal{S}_n^0$

In higher dimensions, a major issue is the fact that the class

$$\mathcal{S}_n := \{ f \in \operatorname{Hol}(\mathbb{B}^n, \mathbb{C}^n) : f(0) = 0, Df(0) = \operatorname{id}, f \text{ univalent} \}$$

of all normalized univalent mappings on $\mathbb{B}^n$ is not compact for any $n \geq 2$. This is easily seen e.g. by considering the noncompact family of shear mappings

$$z = (z_1, \ldots, z_n) \mapsto (z_1 + \alpha z_2^2, z_2, \ldots, z_n), \qquad \alpha \in \mathbb{C},$$

---

[1]Ich vermute, die Gesamtheit dieser Ungleichungen liefere eine vollständige Lösung des Bieberbachschen Koeffizientenproblems [41, p. 363].

which all belong to $\mathcal{S}_n$. In particular, continuous functionals $J : \mathcal{S}_n \to \mathbb{R}$ do not even need to have upper bounds if $n \geq 2$. In order to study extremal problems for univalent functions in higher dimensions, it is therefore necessary to single out a compact subclass of $\mathcal{S}_n$, and one of the most studied classes in this connection is the class $\mathcal{S}_n^0$ of all mappings that admit a so-called parametric representation by means of the Loewner differential equation. It turns out that the classes $\mathcal{S}_n^0$ are compact for each $n \geq 2$ and that $\mathcal{S}_1^0 = \mathcal{S}$.

We now briefly describe the classes $\mathcal{S}_n^0$ using almost standard notation.

Definition 2.1 Let

$$\mathbb{U}_n := \left\{ h \in \mathrm{Hol}(\mathbb{B}^n, \mathbb{C}^n) : h(0) = 0, Dh(0) = -\mathrm{id}, \mathrm{Re}\langle h(z), z \rangle \leq 0 \text{ for all } z \in \mathbb{B}^n \right\}.$$

Here, $\langle \cdot, \cdot \rangle$ denotes the canonical Euclidean inner product of $\mathbb{C}^n$.

It is not difficult to see that a mapping $h \in \mathrm{Hol}(\mathbb{B}^n, \mathbb{C}^n)$ satisfying $h(0) = 0$ and $Dh(0) = -\mathrm{id}$ belongs to $\mathbb{U}_n$ if and only if $\mathrm{Re}\langle -h(z), z \rangle > 0$ for all $z \in \mathbb{B}^n \backslash \{0\}$, see [5, Remark 2.1]. In particular, the set $\mathbb{U}_n$ is exactly the class $-\mathcal{M}$ with $\mathcal{M}$ as defined e.g. in [12, p. 203].

Theorem 2.2 ([12]: Theorem 6.1.39) $\mathbb{U}_n$ *is a compact and convex subset of* $\mathrm{Hol}(\mathbb{B}^n, \mathbb{C}^n)$.

Definition 2.3 Let $\mathbb{R}_0^+ := \{t \in \mathbb{R} : t \geq 0\}$. A *Herglotz vector field in the class* $\mathbb{U}_n$ is a mapping $G : \mathbb{B}^n \times \mathbb{R}_0^+ \to \mathbb{C}^n$ such that

(i) $G(z, \cdot)$ is measurable on $\mathbb{R}_0^+$ for every $z \in \mathbb{B}^n$, and
(ii) $G(\cdot, t) \in \mathbb{U}_n$ for a.e. $t \in \mathbb{R}_0^+$.

Theorem 2.4 (The Loewner Equation) *Let G be a Herglotz vector field in the class $\mathbb{U}_n$. Then for any $z \in \mathbb{B}^n$ there is a unique solution $\mathbb{R}_0^+ \ni t \mapsto \varphi(t, z) \in \mathbb{B}^n$ of the initial value problem*

$$\frac{\partial \varphi}{\partial t}(t, z) = G(\varphi(t, z), t) \quad \text{for a.e. } t \geq 0, \tag{2.1}$$

$$\varphi(0, z) = z.$$

For each $t \geq 0$, the mapping $e^t \varphi(t, \cdot) : \mathbb{B}^n \to \mathbb{C}$ belongs to $\mathcal{S}_n$ and the limit

$$f^G := \lim_{t \to \infty} e^t \varphi(t, \cdot)$$

exists locally uniformly in $\mathbb{B}^n$ and belongs to $\mathcal{S}_n$.

We refer to [12, Thm. 8.1.5] for the proof. The differential equation in (2.1) is the *Loewner equation* (in $\mathbb{C}^n$). It induces a map from the set of all Herglotz vector fields in the class $\mathbb{U}_n$ into the set $\mathcal{S}_n$. The range of this map plays a crucial role in this paper:

Theorem 2.5 *The set*

$$\mathcal{S}_n^0 := \{f^G \in \mathrm{Hol}(\mathbb{B}^n, \mathbb{C}^n) \mid G \ \textit{Herglotz vector field in the class} \ \mathcal{U}_n\}$$

is a compact subset of $\mathrm{Hol}(\mathbb{B}^n, \mathbb{C}^n)$ *for each* $n \in \mathbb{N}$, *and* $\mathcal{S}_1^0 = \mathcal{S}$.

We refer to [12, Corollary 8.3.11] for a proof of the first statement. The fundamental fact that $\mathcal{S}_1^0 = \mathcal{S}$ is a result of Pommerenke [24], see also [25, Chapter 6.1]. We also note that $e^t \varphi(t, \cdot) \in \mathcal{S}_n^0$ for all $t \geq 0$ for every solution to (2.1), see e.g. [39, Lemma 2.6]. The class $\mathcal{S}_n^0$ is exactly the class of mappings in $\mathrm{Hol}(\mathbb{B}^n, \mathbb{C}^n)$ which have a *parametric representation* as introduced by Graham, Hamada and Kohr [13, Definition 1.5], see also [12, 14].

Since the class $\mathcal{S}_n^0$ is compact, we can ask for sharp coefficient bounds as in the one-dimensional case. More precisely, let $f \in \mathcal{S}_n^0$. We write $f = (f_1, \ldots, f_n)$ with $f_j \in \mathrm{Hol}(\mathbb{B}^n, \mathbb{C})$, and consider the coefficient functionals

$$J_\alpha(f) := a_\alpha, \qquad f_1(z) = z_1 + \sum_{\alpha \in \mathbb{N}_0^n, |\alpha| \geq 2} a_\alpha z^\alpha.$$

Here $\alpha = (\alpha_1, \ldots, \alpha_n) \in \mathbb{N}_0^n$ denotes a multi-index and $|\alpha| = \alpha_1 + \ldots + \alpha_n$. In view of the Bieberbach conjecture, it is natural to consider the extremal problem

$$\max_{f \in \mathcal{S}_n^0} \mathrm{Re}\, J_\alpha(f) \tag{2.2}$$

for each multi-index $\alpha \in \mathbb{N}_0^N$ such that $|\alpha| \geq 2$. As in the one-dimensional case, we call a mapping $F \in \mathcal{S}_n^0$ extremal for the functional $\mathrm{Re}\, J_\alpha$ over $\mathcal{S}_n^0$, if

$$\mathrm{Re}\, J_\alpha(F) = \max_{f \in \mathcal{S}_n^0} \mathrm{Re}\, J_\alpha(f).$$

We can now formulate the problem.

Problem 2.6 For a multi-index $\alpha \in \mathbb{N}_0^N$ with $|\alpha| \geq 2$ find a necessary condition for an extremal mapping $F \in \mathcal{S}_n^0$ for the functional $\mathrm{Re}\, J_\alpha$ over $\mathcal{S}_n^0$. For the special case $n = 1$ this condition should reduce to the Schiffer differential equation (1.2). In addition, this necessary condition should be a sufficient condition for extremality under suitable side conditions.

3 Control-Theoretic Interpretation of the Loewner Equation

In this section we give an interpretation of the Loewner equation as an infinite-dimensional *control system* in the Fréchet space $\mathrm{Hol}(\mathbb{B}^n, \mathbb{C}^n)$. For this purpose, we gradually begin to change notation.

Remark 3.1 It seems that Loewner himself was the first who came up with the idea of applying methods from optimal control theory to the Loewner equation. In 1967, his last student G.S. Goodman [11] combined Loewner's theory with the then new Pontryagin Maximum Principle. Since then this approach has been used by many others, see e.g. [1, 9, 10, 26] and in particular the important contributions of D.V. Prokhorov [27–30]. Recent applications of optimal control methods to univalent functions can be found e.g. in [22, 31].

Definition 3.2 (Admissible Controls, Control Set) Let G be a Herglotz vector field in the class $\mathbb{U}_n$. We call the mapping

$$u = u_G : \mathbb{R}_0^+ \to \mathbb{U}_n, \quad t \mapsto G(\cdot, t),$$

an *admissible control* (for the Loewner equation) and denote by $\mathcal{U}_n$ the collection of all admissible controls. The class $\mathbb{U}_n$ is called the *control set* (of the Loewner equation).

Remark 3.3 (Herglotz Vector Fields as $\mathrm{Hol}(\mathbb{B}^n, \mathbb{C}^n)$-*Valued Measurable Controls)* One might think of an admissible control as a measurable mapping defined on the time interval $\mathbb{R}_0^+ = [0, \infty)$ and with values in the control set $\mathbb{U}_n$, which is a subset of the infinite dimensional Fréchet space $\mathrm{Hol}(\mathbb{B}^n, \mathbb{C}^n)$.

We continue to change notation and denote in the sequel the solutions $\varphi(t, \cdot)$ of the Loewner equation (2.1) by φ_t. Note that φ_t always belongs to the composition semigroup

$$\mathbb{S}_n := \left\{ \varphi \in \mathrm{Hol}(\mathbb{B}^n, \mathbb{C}^n) \, : \, \varphi(\mathbb{B}^n) \subseteq \mathbb{B}^n, \, \varphi(0) = 0 \right\}$$

of all holomorphic selfmaps of the unit ball $\mathbb{B}^n$.

Definition 3.4 (Trajectories, State Space) Let $u \in \mathcal{U}_n$ be an admissible control. Denote by G the Herglotz vector field in the class $\mathbb{U}_n$ such that $u = u_G$ and let $t \mapsto \varphi_t$ be the solution of the Loewner equation (2.1) corresponding to G. Then we call the curve

$$x := x_u : \mathbb{R}_0^+ \to \mathbb{S}_n, \quad t \mapsto x_u(t) := \varphi_t,$$

the *trajectory (of the Loewner equation) for* u. The set $\mathbb{S}_n$ is called the *state space* (of the Loewner equation).

Remark 3.5 (Solutions of the Loewner Equation as $\mathbb{S}_n$-*Valued Curves)* One might think of a trajectory $x : \mathbb{R}_0^+ \to \mathbb{S}_n$ of the Loewner equation as an a.e. differentiable (and absolutely continuous) curve defined on the time interval $\mathbb{R}_0^+ = [0, \infty)$ and with values in the state space $\mathbb{S}_n$. The initial point of the curve is the identity map.

In fact, this remark needs clarification. If $x = x_u : \mathbb{R}_0^+ \to \mathbb{S}_n$ is a trajectory of the Loewner equation for $u = u_G$, then for each $z \in \mathbb{B}^n$ there is set $N_z \subseteq \mathbb{R}_0^+$ of measure zero such that the solution $t \mapsto \varphi(t, z)$ of (2.1) is differentiable on $\mathbb{R}_0^+ \setminus N_z$

and the Loewner equation holds for each $t \in \mathbb{R}_0^+ \setminus N_z$. A normal family argument shows that there is in fact a set $N \subseteq \mathbb{R}_0^+$ of measure zero, which does not depend on z, such that the Loewner equation (2.1) holds for each $t \in \mathbb{R}_0^+ \setminus N$ and each $z \in \mathbb{B}^n$. In addition,

$$\dot{x}(t) := \frac{\partial \varphi_t}{\partial t} \in \mathrm{Hol}(\mathbb{B}^n, \mathbb{C}^n) \quad \text{for every } t \in \mathbb{R}_0^+ \setminus N.$$

Note that

$$\dot{x}(t) = \frac{\partial \varphi}{\partial t}(t, \cdot) = G(\varphi(t, \cdot), t) = u(t) \circ x(t), \quad t \in \mathbb{R}_0^+ \setminus N.$$

Conclusion 3.6 (The Loewner Equation as a Control System on $\mathrm{Hol}(\mathbb{B}^n, \mathbb{C}^n)$)
The Loewner equation has the following embarrassingly simple form

$$\boxed{\begin{aligned} \dot{x} &= u \circ x, \\ x(0) &= \mathrm{id} . \end{aligned}} \tag{3.1}$$

This is a control system on the infinite dimensional Fréchet space $\mathrm{Hol}(\mathbb{B}^n, \mathbb{C}^n)$ with state space $\mathbb{S}_n$ and control set $\mathbb{U}_n$.

Remark 3.7 The simple form (3.1) of the Loewner equation is due to E. Schippers [38, Proposition 6].

Remark 3.8 If we define

$$g : \mathbb{S}_n \times \mathbb{U}_n \to \mathrm{Hol}(\mathbb{B}^n, \mathbb{C}^n), \qquad g(x, u) := u \circ x,$$

then the Loewner equation (3.1) takes the "traditional" form of a control system:

$$\begin{aligned} \dot{x}(t) &= g(x(t), u(t)) \quad \text{for a.e. } t \geq 0, \\ x(0) &= \mathrm{id} . \end{aligned} \tag{3.2}$$

Note that the Loewner equation (3.1) is a control system which is linear with respect to the control.

We can now consider the *reachable set for time t* of the Loewner equation, that is, the set $\{x_u(t) : u \in \mathcal{U}_n\}$. However, in view of $e^t x_u(t) \in \mathcal{S}_n^0$ for each $t \geq 0$ and each $u \in \mathcal{U}_n$ (see Theorem 2.5), we strongly prefer to slightly abuse language and call

$$\mathcal{R}(t) := \left\{ e^t x_u(t) : u \in \mathcal{U}_n \right\}, \quad t \in \mathbb{R}_0^+, \quad \text{resp.} \quad \mathcal{R}(\infty) := \left\{ \lim_{t \to \infty} e^t x_u(t) : u \in \mathcal{U}_n \right\}$$

the reachable set for time $t \in \mathbb{R}_+^0 \cup \{\infty\}$ and

$$\mathcal{R} := \bigcup_{t \in [0,\infty]} \mathcal{R}(t)$$

the *overall reachable set* of the Loewner equation.

Using these notions, our considerations can be summarized as follows.

Theorem 3.9 (S_n^0 = **Reachable Set of the Loewner Equation**)

$$S_n^0 = \mathcal{R} = \mathcal{R}(\infty).$$

Remark 3.10 The idea of viewing the classes S_n^0 as reachable sets of the Loewner equation has been pioneered by Prokhorov [28–30] for $n = 1$ and by Graham, Hamada, G. Kohr and M. Kohr [15, 16] for $n > 1$.

With the help of Theorem 3.9, extremal problems over the class S_n^0 can be treated as optimal control problems. For the sake of simplicity, we again consider only extremal problems involving Taylor coefficients of univalent maps.

Definition 3.11 Let $\alpha \in \mathbb{N}_0^N$ be a multi-index with $|\alpha| \geq 2$. An admissible control $u^* \in \mathcal{U}_n$ is called an *optimal control* for the functional $J_\alpha : \mathrm{Hol}(\mathbb{B}^n, \mathbb{C}^n) \to \mathbb{C}$ on S_n^0 if the univalent mapping

$$F_{u^*} := \lim_{t \to \infty} e^t x_{u^*}(t) \in S_n^0$$

is extremal for $\mathrm{Re}\, J_\alpha$ over S_n^0.

Note carefully, that if $F \in S_n^0$ is extremal for $\mathrm{Re}\, J_\alpha$ over S_n^0, then any admissible control $u : [0, \infty) \to \mathbb{U}_n$ which generates F in the sense that $F = F_u$, is an optimal control for J_α. Problem 2.6 can now be stated in control theoretic terms.

Problem 3.12 Let $\alpha \in \mathbb{N}_0^N$ be a multi-index with $|\alpha| \geq 2$. Find a necessary condition for an optimal control for the functional $J_\alpha : \mathrm{Hol}(\mathbb{B}^n, \mathbb{C}^n) \to \mathbb{C}$. In addition, this necessary condition should be a sufficient condition for extremality under suitable side conditions.

4 The Pontryagin Maximum Principle for the Class S_n^0

The standard necessary condition for an optimal control is provided by the

Pontryagin Maximum Principle,

see [2]. We first consider the standard finite dimensional version.

Remark 4.1 (Prelude to the Pontryagin Maximum Principle) Let V be a finite dimensional complex vector space. Suppose that the "state space" S is an open subset of V and that the "control set" U is a compact subset of V. Let $g : S \times U \to V$ be a C^1-function and let $x_0 \in S$ be a fixed "initial" state.

(I) *The control system*
Consider

$$\dot{x}(t) = g(x(t), u(t)) \quad \text{for a.e. } t \geq 0,$$
$$x(0) = x_0,$$

(4.1)

and call a measurable control $u : \mathbb{R}_0^+ \to U$ *admissible* (for the system (4.1)), if the solution $x = x_u$ to (4.1) exists for all a.e. $t \in \mathbb{R}_0^+$.

(II) *The optimal control problem*
Let $J \in V^*$, that is, let $J : V \to \mathbb{C}$ be a $\mathbb{C}$-linear (continuous) functional and let $T > 0$ be fixed. We call an admissible control $u^* : \mathbb{R}_0^+ \to U$ an *optimal control* for the functional $J : V \to \mathbb{C}$ on S and time $T > 0$, if

$$\operatorname{Re} J(x_{u^*}(T)) = \max \left\{ \operatorname{Re} J(x_u(T)) : u : \mathbb{R}_0^+ \to U \text{ admissible} \right\}. \quad (4.2)$$

(III) *The adjoint equation*
Suppose that $u : \mathbb{R}_0^+ \to U$ is an admissible control and $x = x_u : \mathbb{R}_0^+ \to S$ is the solution to (4.1). Then the linear (matrix) equation

$$\dot{\Phi}(t) = -\Phi(t) D_x g(x(t), u(t)),$$
$$\Phi(T) = \mathrm{id},$$

(4.3)

is called the *adjoint equation* along (x, u) at T. Note that $\Phi(t) \in \mathcal{L}(V)$, the set of linear (continuous) endomorphisms of V.

(IV) *The Hamiltonian*
The function $\mathcal{H} : S \times \mathcal{L}(V) \times U \to \mathbb{C}$ defined by

$$\mathcal{H}(x, \Phi, u) := J(\Phi \cdot g(x, u))$$

is called the *complex Hamiltonian* for the extremal problem (4.2). Note that if we denote the transpose of a map $\Phi \in \mathcal{L}(V)$ by Φ_*, then

$$\mathcal{H}(x, \Phi, u) = \Phi_*(J) g(x, u)$$

We can now state the Pontryagin Maximum Principle:

Theorem 4.2 (Pontryagin Maximum Principle) *Let V be a finite dimensional complex vector space and $J \in V^*$. Suppose that $u^* : \mathbb{R}_0^+ \to U$ is an optimal control for $J : V \to \mathbb{C}$ on S and time $T > 0$. Denote by $\Phi^* : \mathbb{R}_0^+ \to \mathcal{L}(V)$ the* *solution to the adjoint equation (4.3) along* (x_{u^*}, u^*) *at T, then*

$$\operatorname{Re} \mathcal{H}(x_{u^*}(t), \Phi^*(t), u^*(t)) = \max_{u \in U} \operatorname{Re} \mathcal{H}(x_{u^*}(t), \Phi^*(t), u) \quad \text{for a.e. } t \in [0, T].$$

The conclusion of Theorem 4.2 means that for a.e. $t \in \mathbb{R}_0^+$ the value $u^*(t)$ of the optimal control u^* provides a maximum for the function

$$u \mapsto \operatorname{Re} \Phi^*(t)_*(J)g(x_*(t), u)$$

over the control set U. For a proof of Theorem 4.2 we refer to any textbook on optimal control theory, see e.g. [43, p. 152 ff.].

Remark 4.3 (On the Definition of the Hamiltonian/The Costate Equation) Our definition of the Hamiltonian is slightly nonstandard. However, it is easy to see its relation to the standard Hamiltonian formalism. Note that for $V = \mathbb{C}^n$ we can identify the dual space V^* with V and we can write the $\mathbb{C}$-linear functional $J : V \to \mathbb{C}$ as $J(y) = \overline{\eta}^T y$ for some vector $\eta \in V$. Hence, if $\Phi : \mathbb{R}_0^+ \to \mathcal{L}(V)$ is the solution of the adjoint equation (4.3) along (x, u) at T, then

$$\Psi := \overline{\eta}^T \Phi = J(\Phi \cdot) = \Phi_*(J) : \mathbb{R}_+^0 \to V^*$$

is the solution to the so-called *costate equation*

$$\dot{\Psi}(t) = -\Psi(t) D_x g(x(t), u(t)),$$
$$\Psi(T) = \overline{\eta}^T. \tag{4.4}$$

Therefore, the *standard* complex Hamiltonian for the extremal problem (4.2),

$$H : S \times V^* \times U \to \mathbb{C}, \quad H(x, \Psi, u) := \Psi^T g(x, u), \tag{4.5}$$

see [21, 29], is related to (our) complex Hamiltonian by

$$\mathcal{H}(x, \Phi, u) = H(x, \Psi, u), \qquad \Psi = \overline{\eta}^T \Phi.$$

We next apply the Pontryagin machinery as outlined above to the abstract Loewner equation (3.1), so we replace the finite dimensional complex vector space V, the state space $S \subseteq V$, the control set $U \subseteq V$ and the control system $g : S \times U \to V$ by

$$V = \operatorname{Hol}(\mathbb{B}^n, \mathbb{C}^n), \quad S = \mathbb{S}_n, \quad U = \mathbb{U}_n, \quad g(x, u) = x \circ u.$$

Remark 4.4 As we shall see, our slightly nonstandard definition of the Hamiltonian proves itself as user-friendly. The main reason for this is that the dual $\operatorname{Hol}(\mathbb{B}^n, \mathbb{C}^n)^*$

Table 1 Loewner equation as a control system on $\mathrm{Hol}(\mathbb{B}^n, \mathbb{C}^n)$

	Control system on $\mathbb{C}^n$	Loewner equation
State space	$S \subseteq \mathbb{C}^n$	$\mathbb{S}_n \subseteq \mathrm{Hol}(\mathbb{B}^n, \mathbb{C}^n)$
Control set	$U \subseteq \mathbb{C}^n$	$\mathbb{U}_n \subseteq \mathrm{Hol}(\mathbb{B}^n, \mathbb{C}^n)$
Controls	$u : \mathbb{R}_0^+ \to U$	$u : \mathbb{R}_0^+ \to \mathbb{U}_n$
Trajectories	$x : \mathbb{R}_0^+ \to S$	$x : \mathbb{R}_0^+ \to \mathbb{S}_n$
Equation	$\dot{x} = g(x, u)$	$\dot{x} = u \circ x$

of $\mathrm{Hol}(\mathbb{B}^n, \mathbb{C}^n)$ can no longer be identified with $\mathrm{Hol}(\mathbb{B}^n, \mathbb{C}^n)$, and in fact the topological structure of $\mathrm{Hol}(\mathbb{B}^n, \mathbb{C}^n)^*$ is fairly complicated, see [6, 17, 18, 42]. However, it will turn out that in case of the Loewner equation, the solutions of the associated adjoint equation actually live in the "decent" vector space $\mathrm{Hol}(\mathbb{B}^n, \mathbb{C}^{n \times n})$ of all holomorphic maps from $\mathbb{B}^n$ into $\mathbb{C}^{n \times n}$. Note that each $A \in \mathrm{Hol}(\mathbb{B}^n, \mathbb{C}^{n \times n})$ gives rise to the continuous linear operator $\mathcal{A} \in \mathcal{L}(\mathrm{Hol}(\mathbb{B}^n, \mathbb{C}^n))$ defined by

$$\mathcal{A}f := A(\cdot)f, \qquad f \in \mathrm{Hol}(\mathbb{B}^n, \mathbb{C}^n).$$

We proceed in a purely formal way in order to emphasize the analogy with the finite dimensional case, but wish to point out that the following formal considerations can be made rigorous using the elegant Fréchet space calculus developed by R. Hamilton [19]. We start with the adjoint equation for the Loewner ODE (Table 1).

Remark 4.5 (The Adjoint Equation of the Loewner ODE) Recall the Loewner ODE in abstract form

$$\dot{x} = u \circ x.$$

Taking the "functional" derivative of the mapping on the right-hand side,

$$g : \mathbb{S}_n \times \mathbb{U}_n \to \mathrm{Hol}(\mathbb{B}^n, \mathbb{C}^n), \qquad g(x, u) := u \circ x,$$

with respect to $x \in \mathbb{S}_n$, we get

$$\frac{\partial(u \circ x)}{\partial x} = Du \circ x.$$

Here, $Du \in \mathrm{Hol}(\mathbb{B}^n, \mathbb{C}^{n \times n})$ is the total derivative of $u \in \mathrm{Hol}(\mathbb{B}^n, \mathbb{C}^n)$ and hence $Du \circ x \in \mathrm{Hol}(\mathbb{B}^n, \mathbb{C}^n)$. Therefore, the adjoint equation of the Loewner ODE is

$$\dot{\Phi}(t) = -\Phi(t) \cdot Du(t) \circ x(t), \tag{4.6}$$

$$\Phi(T) = \mathrm{id}.$$

Recall that for each $t \in \mathbb{R}_0^+$ the map $\Phi(t)$ is a continuous linear operator on $\mathrm{Hol}(\mathbb{B}^n, \mathbb{C}^n)$. It is now easy to find the solution of (4.6) in terms of Dx. In fact, just take the total derivative of $\overset{\bullet}{x} = u \circ x \in \mathrm{Hol}(\mathbb{B}^n, \mathbb{C}^n)$ and get

$$\overset{\bullet}{Dx} = Du \circ x \cdot Dx.$$

Hence using $Dx \cdot (Dx)^{-1} = \mathrm{id}$, we see that

$$\overset{\bullet}{(Dx)^{-1}} = -(Dx)^{-1} Du \circ x.$$

This means that the solution to the adjoint equation (4.6) along (x, u) at T is simply

$$t \mapsto \Phi(t) = Dx(T) \cdot (Dx(t))^{-1} \in \mathrm{Hol}(\mathbb{B}^n, \mathbb{C}^{n \times n}) \subseteq \mathcal{L}(\mathrm{Hol}(\mathbb{B}^n, \mathbb{C}^n)),$$

and the corresponding Hamiltonian of the Loewner equation (3.1) for the extremal problem (4.2) has the form

$$\mathcal{H}(x, \Phi, u) = J\left(Dx(T) \cdot (Dx)^{-1} \cdot u \circ x\right), \quad u \in \mathbb{U}_n.$$

Definition 4.6 Let $J \in \mathrm{Hol}(\mathbb{B}^n, \mathbb{C}^n)^*$ and $T > 0$. Suppose that $u^* \in \mathcal{U}_n$. Then we define for each $t \geq 0$,

$$L_t \in \mathrm{Hol}(\mathbb{B}^n, \mathbb{C}^n)^*, \qquad L_t(u) := J\left(Dx_{u^*}(T) \cdot (Dx_{u^*}(t))^{-1} \cdot u \circ x_{u^*}(t)\right).$$

Now we can state the analogue of Theorem 4.2 for the Loewner equation and general linear functionals $J \in \mathrm{Hol}(\mathbb{B}^n, \mathbb{C}^n)^*$.

Theorem 4.7 (Pontryagin Maximum Principle for the Loewner equation) *Let* $J \in \mathrm{Hol}(\mathbb{B}^n, \mathbb{C}^n)^*$. *Suppose that* $u^* : \mathbb{R}_0^+ \to \mathbb{U}_n$ *is an optimal control for* $J : \mathrm{Hol}(\mathbb{B}^n, \mathbb{C}^n) \to \mathbb{C}$ *on* $\mathbb{S}_n$ *and time* $T > 0$. *Then*

$$\mathrm{Re}\, L_t(u^*(t)) = \max_{u \in \mathbb{U}_n} \mathrm{Re}\, L_t(u) \quad \text{for a.e. } t \in [0, T].$$

It is now a short step to a statement of the Pontryagin Maximum Principle for the class $\mathcal{S}_n^0$ and coefficient functionals J_α.

Theorem 4.8 (Pontryagin Maximum Principle for the class $\mathcal{S}_n^0$) *Let* $\alpha \in \mathbb{N}_0^N$ *with* $|\alpha| \geq 2$ *and let* $F \in \mathcal{S}_n^0$ *be an extremal mapping for* $\mathrm{Re}\, J_\alpha$ *over* $\mathcal{S}_n^0$. *Suppose that* G *is a Herglotz vector field in the class* $\mathbb{U}_n$ *such that* $F = f^G$. *Denote by* $\varphi_t : \mathbb{R}_0^+ \to \mathbb{S}_n$ *the solution to the Loewner equation (2.1). For each* $t \geq 0$ *define*

$$L_t \in \mathrm{Hol}(\mathbb{B}^n, \mathbb{C}^n)^*, \quad h \mapsto L_t(h) := J_\alpha\left(DF \cdot [D\varphi_t]^{-1} \cdot h(\varphi_t)\right). \tag{4.7}$$

Then for a.e. $t \geq 0$,

$$\operatorname{Re} L_t(G(\cdot, t)) = \max_{h \in \mathbb{U}_n} \operatorname{Re} L_t(h).$$ (4.8)

See [33] for a rigorous proof of Theorem 4.8.

Conclusion 4.9 Let $\alpha \in \mathbb{N}_0^N$ with $|\alpha| \geq 2$ and let $F \in \mathcal{S}_n^0$. Suppose that G is a Herglotz vector field in the class $\mathbb{U}_n$ such that $F = f^G$. Then the condition that

> $G(\cdot, t)$ maximizes $\operatorname{Re} L_t$ over $\mathbb{U}_n$ for a.e. $t \geq 0$

is necessary for F being extremal for $\operatorname{Re} J_\alpha$ over $\mathcal{S}_n^0$.

5 The Schiffer Differential Equation and Pontryagin's Maximum Principle

We now briefly describe the intimate relation between Schiffer's differential equation and Pontryagin's Maximum Principle for the case $\mathcal{S}_1^0 = \mathcal{S}$. For the sake of simplicity, we restrict again to the case of the N-th coefficient functional $J_N : \mathcal{S} \to \mathbb{C}$, $J_N(f) := f^{(N)}(0)/N!$.

Theorem 5.1 *Let $F \in \mathcal{S}$ and suppose that G is a Herglotz vector field in the class $\mathbb{U}_1$ such that $F = f^G$. Denote by $\varphi_t : \mathbb{R}_0^+ \to \mathbb{S}_1$ the solution to the Loewner equation (2.1). For each $t \geq 0$ define*

$$L_t \in \operatorname{Hol}(\mathbb{D}, \mathbb{C}), \quad h \mapsto L_t(h) := J_N\left(F' \cdot (\varphi_t')^{-1} \cdot h(\varphi_t)\right).$$ (5.1)

Then the following conditions are equivalent.

(a) F is a solution of the Schiffer differential equation (1.2), that is,

$$\left[\frac{zF'(z)}{F(z)}\right]^2 P_N\left(\frac{1}{F(z)}\right) = R_N(z),$$ (5.2)

with P_N resp. R_N defined by (1.1) resp. (1.3), and the positivity condition (1.4) holds.

(b) For a.e. $t \geq 0$,

$$\max_{h \in \mathbb{U}_1} \operatorname{Re} L_t(h) = \operatorname{Re} L_t(G(\cdot, t)).$$ (5.3)

Remark 5.2 (Schiffer's Equation = Pontryagin's Maximum Principle) Note that *under the assumption that F is extremal for the functional J_N over S we have:*

 (i) Condition (a) = Conclusion of Schiffer's Theorem 1.2;
(ii) Condition (b) = Conclusion of Pontryagin's Maximum Principle
                     (Theorem 4.8 for the case $n = 1$).

Hence Theorem 5.1 says that the two necessary conditions for F being extremal for the functional J_N over S provided by Schiffer's theorem (Condition (a)) and by Pontryagin's Maximum Principle (Condition (b)) are in fact *equivalent*.

A few more words are in order.

Remark 5.3 The implication (a) $\Longrightarrow$ (b) of Theorem 5.1 can in fact be traced back to the work of Schaeffer, Schiffer and Spencer, see [34, 35, 37]. Consequently, a version of Pontryagin's Maximum Principle for a certain control system, namely the Loewner equation for $n = 1$, has been known more than a decade before the discovery of the maximum principle by Pontryagin and his coauthors in 1956. In fact, Leung [23] speaks in this regard of the *Schiffer–Pontryagin Maximum Principle*. The converse implication (b) $\Longrightarrow$ (a) seems to lie deeper, see [11, 26, 32].

Conclusion 5.4 (Extending Schiffer's Theorem to Higher Dimensions) Theorem 4.8 generalizes Schiffer's Theorem 1.2 from the class $S = S_1^0$ to any of the classes S_n^0.

6   Teichmüller's Coefficient Theorem Without Quadratic Differentials

Using Theorem 5.1 it is now easy to give a formulation of Teichmüller's Coefficient Theorem solely in terms of the Loewner differential equation.

Theorem 6.1 *Let $F \in S$ and suppose that G is a Herglotz vector field in the class $\mathbb{U}_1$ such that $F = f^G$. Denote by $\varphi_t : \mathbb{R}_0^+ \to \mathbb{S}_1$ the solution to the Loewner equation (2.1). For each $t \geq 0$ define $L_t \in \mathrm{Hol}(\mathbb{D}, \mathbb{C})$ by (5.1). If for a.e. $t \geq 0$,*

$$\max_{h \in \mathbb{U}_1} \mathrm{Re}\, L_t(h) = \mathrm{Re}\, L_t(G(\cdot, t)), \tag{6.1}$$

then

$$\mathrm{Re}\, J_N(f) \leq \mathrm{Re}\, J_N(F)$$

for any

$$f \in S(A_2, \ldots, A_{N-1}).$$

Proof By Theorem 5.1 and Remark 1.3 (d) this exactly is Teichmüller's Coefficient Theorem 1.4. □

Note that in view of Pontryagin's Maximum Principle, condition (6.1) is necessary for $F \in S$ being an extremal function for J_N over S. Theorem 6.1 simply says that this condition is also sufficient for extremality under a suitable side condition.

Conclusion 6.2 Let $F(z) = z + \sum_{k=2}^{\infty} A_k z^k \in S$ and suppose that G is a Herglotz vector field in the class $\mathbb{U}_1$ such that $F = f^G$. For each $t \geq 0$ define $L_t \in \mathrm{Hol}(\mathbb{D}, \mathbb{C})$ by (5.1). Then the condition that

$$\boxed{G(\cdot, t) \text{ maximizes } \mathrm{Re}\, L_t \text{ over } \mathbb{U}_1 \text{ for a.e. } t \geq 0}$$

is

(a) necessary for F being extremal for $\mathrm{Re}\, J_N$ over S, and
(b) sufficient for F being extremal for $\mathrm{Re}\, J_N$ over $S(A_2, \dots, A_{N-1})$.

Problem 6.3 Find a proof of Theorem 6.1 using only the Loewner differential equation. We note that the standard method in control theory for obtaining sufficient conditions for optimal control functions makes use of Bellman functions. See [29] for some application of Bellman functions to the Loewner equation.

Problem 6.4 Let $\alpha \in \mathbb{N}_0^N$ be a multi-index with $|\alpha| \geq 2$, $F \in S_n^0$ and suppose that G is a Herglotz vector field in the class $\mathbb{U}_n$ such that $F = f^G$. Denote by $\varphi_t : \mathbb{R}_0^+ \to S_n$ the solution to the Loewner equation (2.1). For each $t \geq 0$ define $L_t \in \mathrm{Hol}(\mathbb{B}^n, \mathbb{C}^n)$ by (4.7). Assume that for a.e. $t \geq 0$,

$$\max_{h \in \mathbb{U}_n} \mathrm{Re}\, L_t(h) = \mathrm{Re}\, L_t(G(\cdot, t)). \tag{6.2}$$

Is there a subset S_α of S_n^0 such that

$$\mathrm{Re}\, J_\alpha(f) \leq \mathrm{Re}\, J_\alpha(F)$$

for any $f \in S_\alpha$? In other words, is the necessary condition for $F \in S_n^0$ being an extremal function for J_α over S_n^0 provided by Pontryagin's Maximum Principle also sufficient under a suitable side condition? For $n = 1$ the answer is "Yes" by Theorem 6.1, which we have seen is equivalent to Teichmüller's Coefficient Theorem 1.4. An affirmative answer for $n > 1$ would therefore provide an extension of Teichmüller's Coefficient Theorem to higher dimensions.

References

1. Aleksandrov, I.: Parametric Continuations in the Theory of Univalent Functions (Parametricheskie prodolzhenya teorii odnolistnykh funktsij), p. 343. Nauka, Moskva (1976)
2. Boltyanskiĭ, V.G., Gamkrelidze, R.V., Pontryagin, L.S.: On the theory of optimal processes. Dokl. Akad. Nauk SSSR (N.S.) **110**, 7–10 (1956)
3. Bracci, F., Roth, O.: Support points and the Bieberbach conjecture in higher dimension. https://arxiv.org/abs/1603.01532
4. Bracci, F., Contreras, M.D., Díaz-Madrigal, S., Vasil'ev, A.: Classical and stochastic Löwner-Kufarev equations. In: Harmonic and Complex Analysis and Its Applications, pp. 39–134. Birkhäuser/Springer, Cham (2014)
5. Bracci, F., Graham, I., Hamada, H., Kohr, G.: Variation of Loewner chains, extreme and support points in the class S^0 in higher dimensions. Constr. Approx. **43**(2), 231–251 (2016)
6. Caccioppoli, R.: Sui funzionali lineari nel campo delle funzioni analitiche. Atti Accad. Naz. Lincei, Rend., VI. Ser. **13**, 263–266 (1931)
7. de Branges, L.: A proof of the Bieberbach conjecture. Acta Math. **154**, 137–152 (1985)
8. Duren, P.L.: Univalent functions. Grundlehren der Mathematischen Wissenschaften [Fundamental Principles of Mathematical Sciences], vol. 259. Springer, New York (1983)
9. Friedland, S., Schiffer, M.: Global results in control theory with applications to univalent functions. Bull. Am. Math. Soc. **82**(6), 913–915 (1976)
10. Friedland, S., Schiffer, M.: On coefficient regions of univalent functions. J. Analyse Math. **31**, 125–168 (1977)
11. Goodman, G.S.: Univalent functions and optimal control. Ph.D. Thesis, Stanford University. ProQuest LLC, Ann Arbor, MI (1967)
12. Graham, I., Kohr, G.: Geometric function theory in one and higher dimensions. In: Monographs and Textbooks in Pure and Applied Mathematics, vol. 255. Marcel Dekker, New York (2003)
13. Graham, I., Hamada, H., Kohr, G.: Parametric representation of univalent mappings in several complex variables. Canad. J. Math. **54**(2), 324–351 (2002)
14. Graham, I., Hamada, H., Kohr, G., Kohr, M.: Parametric representation and asymptotic starlikeness in $\mathbb{C}^n$. Proc. Am. Math. Soc. **136**(11), 3963–3973 (2008)
15. Graham, I., Hamada, H., Kohr, G., Kohr, M.: Extreme points, support points and the Loewner variation in several complex variables. Sci. China Math. **55**(7), 1353–1366 (2012)
16. Graham, I., Hamada, H., Kohr, G., Kohr, M.: Extremal properties associated with univalent subordination chains in $\mathbb{C}^n$. Math. Ann. **359**(1-2), 61–99 (2014)
17. Grothendieck, A.: Sur certains espaces de fonctions holomorphes. I. J. Reine Angew. Math. **192**, 35–64 (1953)
18. Grothendieck, A.: Sur certains espaces de fonctions holomorphes. II. J. Reine Angew. Math. **192**, 77–95 (1953)
19. Hamilton, R.S.: The inverse function theorem of Nash and Moser. Bull. Am. Math. Soc. (N.S.) **7**(1), 65–222 (1982)
20. Jenkins, J.A.: Univalent functions and conformal mapping. Ergebnisse der Mathematik und ihrer Grenzgebiete. Neue Folge, Heft 18. Reihe: Moderne Funktionentheorie. Springer, Berlin (1958)
21. Jurdjevic, V.: Geometric Control Theory. Cambridge Studies in Advanced Mathematics, vol. 52. Cambridge University Press, Cambridge (1997)
22. Koch, J., Schleißinger, S.: Value ranges of univalent self-mappings of the unit disc. J. Math. Anal. Appl. **433**(2), 1772–1789 (2016)
23. Leung, Y.J.: Notes on Loewner differential equations. In: Topics in Complex Analysis (Fairfield, Conn., 1983). Contemporary Mathematics, vol. 38, pp. 1–11. American Mathematical Society, Providence, RI (1985)
24. Pommerenke, C.: Über die Subordination analytischer Funktionen. J. Reine Angew. Math. **218**, 159–173 (1965)

25. Pommerenke, C.: Univalent functions. Vandenhoeck & Ruprecht, Göttingen, 1975. With a chapter on quadratic differentials by Gerd Jensen, Studia Mathematica/Mathematische Lehrbücher, Band XXV.
26. Popov, V.: L.S. Pontryagin's maximum principle in the theory of univalent functions. Sov. Math. Dokl. **10**, 1161–1164 (1969)
27. Prokhorov, D.: The method of optimal control in an extremal problem on a class of univalent functions. Sov. Math. Dokl. **29**, 301–303 (1984)
28. Prokhorov, D.V.: Sets of values of systems of functionals in classes of univalent functions. Mat. Sb. **181**(12), 1659–1677 (1990)
29. Prokhorov, D.V.: Reachable Set Methods in Extremal Problems for Univalent Functions. Saratov University Publishing House, Saratov (1993)
30. Prokhorov, D.V.: Bounded univalent functions. In: Handbook of Complex Analysis: Geometric Function Theory, vol. 1, pp. 207–228. North-Holland, Amsterdam (2002)
31. Prokhorov, D., Samsonova, K.: Value range of solutions to the chordal Loewner equation. J. Math. Anal. Appl. **428**(2), 910–919 (2015)
32. Roth, O.: Pontryagin's maximum principle in geometric function theory. Complex Var. Theory Appl. **41**(4), 391–426 (2000)
33. Roth, O.: Pontryagin's maximum principle for the Loewner equation in higher dimensions. Canad. J. Math. **67**(4), 942–960 (2015)
34. Schaeffer, A.C., Spencer, D.C.: Coefficient regions for Schlicht functions. With a Chapter on the Region of the Derivative of a Schlicht Function by Arthur Grad, vol. 35. American Mathematical Society, New York, NY (1950)
35. Schaeffer, A.C., Schiffer, M., Spencer, D.C.: The coefficient regions of Schlicht functions. Duke Math. J. **16**, 493–527 (1949)
36. Schiffer, M.: A method of variation within the family of simple functions. Proc. Lond. Math. Soc. (2) **44**, 432–449 (1938)
37. Schiffer, M.: Sur l'équation différentielle de M. Löwner. C. R. Acad. Sci. Paris **221**, 369–371 (1945)
38. Schippers, E.: The power matrix, coadjoint action and quadratic differentials. J. Anal. Math. **98**, 249–277 (2006)
39. Schleissinger, S.: On support points of the class $S^0(B^n)$. Proc. Am. Math. Soc. **142**(11), 3881–3887 (2014)
40. Strebel, K.: Quadratic differentials. Ergebnisse der Mathematik und ihrer Grenzgebiete. 3. Folge, Band 5, p. 184. Berlin etc. Springer. XII (1984)
41. Teichmüller, O.: Ungleichungen zwischen den Koeffizienten schlichter Funktionen.Sitzungsber. Preuß. Akad. Wiss., Phys.-Math. Kl. **1938**, 363–375 (1938)
42. Toeplitz, O.: Die linearen vollkommenen Räume der Funktionentheorie. Comment. Math. Helv. **23**, 222–242 (1949)
43. Zabczyk, J.: Mathematical Control Theory: An Introduction. Systems & Control: Foundations & Applications. Birkhäuser, Boston, MA (1992)

Open Problems Related to a Herglotz-Type Formula for Vector-Valued Mappings

Jerry R. Muir, Jr.

Abstract We review a recently found Herglotz-type formula that represents mappings in the class $\mathcal{M}$, the generalization of the Carathéodory class to the open unit ball $\mathbb{B}$ of $\mathbb{C}^n$, as integrals of a fixed kernel with respect to a family of probability measures on $\partial \mathbb{B}$. Since not every probability measure arises in the representation, there are resulting questions regarding the characterization of the family of representing measures. In a manner similar to how the one-variable Herglotz formula facilitates a representation of convex mappings of the disk, we apply the new transformation to convex mappings of $\mathbb{B}$ and present some additional open questions. As part of this application, we present a proof of an unpublished result of T.J. Suffridge that generalizes a classical result of Marx and Strohhäcker for convex mappings of the disk.

Keywords Herglotz formula · Carathéodory class · Convex mappings

2010 Mathematics Subject Classification Primary 32H02, 32A26; Secondary 30C45

1 Introduction

Central to a variety of concepts in the subject of geometric function theory in higher dimensions is the class $\mathcal{M}$. While this family can be defined on the open unit ball of any complex Banach space, we will restrict our discussion to the open Euclidean unit ball $\mathbb{B} \subseteq \mathbb{C}^n$. In this case, we have

$$\mathcal{M} = \{f \in H(\mathbb{B}, \mathbb{C}^n) : f(0) = 0, \ Df(0) = I, \text{ and } \mathrm{Re}\langle f(z), z \rangle > 0 \text{ if } z \in \mathbb{B} \setminus \{0\}\},$$

J. R. Muir (✉)

Department of Mathematics, The University of Scranton, Scranton, PA, USA

e-mail: jerry.muir@scranton.edu

© Springer International Publishing AG, part of Springer Nature 2017

F. Bracci (ed.), *Geometric Function Theory in Higher Dimension*,

Springer INdAM Series 26, https://doi.org/10.1007/978-3-319-73126-1_8

where $H(\mathbb{B}, \mathbb{C}^n)$ denotes the space of holomorphic mappings from $\mathbb{B}$ into $\mathbb{C}^n$, Df is the Fréchet derivative of an $f \in H(\mathbb{B}, \mathbb{C}^n)$, I is the identity operator on $\mathbb{C}^n$, and $\langle \cdot, \cdot \rangle$ is the Hermitian inner product in $\mathbb{C}^n$. We will also denote the Euclidean norm in $\mathbb{C}^n$ by $\| \cdot \|$.

The class $\mathcal{M}$ is key for providing analytic necessary and sufficient conditions for a locally biholomorphic mapping $f \in H(\mathbb{B}, \mathbb{C}^n)$ to map $\mathbb{B}$ biholomorphically onto a domain in $\mathbb{C}^n$ with a particular geometric property. For example, if such an f satisfies $f(0) = 0$, then f is a starlike mapping (i.e., f is biholomorphic and $f(\mathbb{B})$ is a starlike domain with respect to 0) if and only if $g \in \mathcal{M}$, where $g(z) = Df(z)^{-1}f(z)$, $z \in \mathbb{B}$. Additionally, the class $\mathcal{M}$ is fundamental to the Loewner theory on $\mathbb{B}$. Indeed, the classical Loewner differential equation on $\mathbb{B}$ is

$$\frac{\partial}{\partial t} f_t(z) = Df_t(z)h(z, t), \qquad z \in \mathbb{B}, \text{ a.e. } t \geq 0,$$

where $h: \mathbb{B} \times [0, \infty) \to \mathbb{C}^n$ is such that $h(\cdot, t) \in \mathcal{M}$ for all $t \geq 0$ and $h(z, \cdot)$ is measurable for all $z \in \mathbb{B}$. Solutions $\{f_t\}_{t \geq 0} \subseteq H(\mathbb{B}, \mathbb{C}^n)$ that satisfy some additional requirements are Loewner chains on $\mathbb{B}$. See [2] for a good presentation on all such topics.

A common theme in this flavor of geometric function theory, many problems considered on $\mathbb{B}$ are motivated by the classical theory on the open unit disk $\mathbb{D} \subseteq \mathbb{C}$, but the higher-dimensional setting introduces complications that must be addressed by a novel, more intricate theory. In this case, the class $\mathcal{M}$ is seen to be a generalization of the Carathéodory class $\mathcal{P}$, consisting of all $f \in H(\mathbb{D}, \mathbb{C})$ such that $f(0) = 1$ and $\operatorname{Re} f(z) > 0$ for $z \in \mathbb{D}$. Indeed, if $g(z) = zf(z)$ for some $f \in H(\mathbb{D}, \mathbb{C})$, then $g \in \mathcal{M}$ if and only if $f \in \mathcal{P}$.

Of course, the theory of geometric mappings and the Loewner theory on $\mathbb{D}$ are closely related to the class $\mathcal{P}$. These relationships then benefit greatly from the following well-known Herglotz representation for $\mathcal{P}$. Let $P(\partial \mathbb{D})$ be the set of Borel probability measures on $\partial \mathbb{D}$. As a subset of the dual of the Banach space of complex-valued continuous functions on $\partial \mathbb{D}$, we may consider the weak-$*$ topology on $P(\partial \mathbb{D})$.

Theorem 1.1 *Let $f: \mathbb{D} \to \mathbb{C}$. Then $f \in \mathcal{P}$ if and only if there exists $\mu = \mu_f \in P(\partial \mathbb{D})$ such that*

$$f(z) = \int_{\partial \mathbb{D}} \frac{u + z}{u - z} \, d\mu(u), \qquad z \in \mathbb{D}. \tag{1.1}$$

The mapping $f \mapsto \mu_f$ is an affine homeomorphism of $\mathcal{P}$ onto $P(\partial \mathbb{D})$ when $\mathcal{P}$ and $P(\partial \mathbb{D})$ are respectively endowed with the topology of uniform convergence on compact subsets of $\mathbb{D}$ and the weak-$*$ topology.

The implications of this representation are tremendous. Because $\mathcal{P}$ and $P(\partial \mathbb{D})$ are compact convex sets in their respective topologies, their extreme points are in correspondence. Thus the fact that the Dirac measures δ_α for $\alpha \in \partial \mathbb{D}$ are the extreme points of $P(\partial\mathbb{D})$ implies that the right half-plane mappings $z \mapsto (\alpha + z)/(\alpha - z)$, $z \in \mathbb{D}$, are the extreme points of $\mathcal{P}$. Brickman et al. [1] used the Herglotz representation to express families of geometric mappings on $\mathbb{D}$ (convex mappings, starlike mappings, etc.) as integrals and subsequently identify the extreme points of their closed convex hulls.

Recently [4], the author provided a Herglotz-type representation for the class $\mathcal{M}$. The purpose of this note is twofold. First, we will review the nature of this new Herglotz representation and provide some open questions resulting from it. Then we will consider the application of the representation to the family of convex mappings of $\mathbb{B}$. This will include providing a proof of an unpublished result of Ted J. Suffridge that generalizes a theorem of Marx and Strohhäcker from $\mathbb{D}$ to $\mathbb{B}$ and will result in more open questions.

2 The Herglotz-Type Representation on $\mathbb{B}$

In order to state the representation, we first develop some notation. Let $\mathbb{S} = \partial\mathbb{B}$ denote the unit sphere, and write $C(\mathbb{S})$ for the space of complex-valued continuous functions on $\mathbb{S}$. The space $M(\mathbb{S})$ of regular complex Borel measures on $\mathbb{S}$ is isometrically isomorphic to the dual space $C(\mathbb{S})^*$ and inherits its weak-$*$ topology. Let $P(\mathbb{S}) \subseteq M(\mathbb{S})$ be the set of probability measures, and let $\sigma \in P(\mathbb{S})$ be its unique rotation-invariant element; i.e., σ is the normalized Lebesgue surface area measure.

Let $\mathbb{N}_0^n$ (where $\mathbb{N}_0 = \mathbb{N} \cup \{0\}$) denote the set of multi-indices and E be the span of (i.e., all finite linear combinations of) all monomials of the form

$$u \mapsto \bar{u}_k u^\alpha, \qquad u \mapsto u_k \bar{u}^\alpha, \qquad u \in \mathbb{S},$$

for $k \in \{1, \ldots, n\}$ and $\alpha \in \mathbb{N}_0^n \setminus \{0\}$. Here, $u^\alpha = \prod_{k=1}^n u_k^{\alpha_k}$. We now define

$$\hat{E} = \{\mu \in M(\mathbb{S}) : \ d\mu = p \, d\sigma \text{ for some } p \in E\}.$$

Finally, set $P_E(\mathbb{S})$ to be the intersection of $P(\mathbb{S})$ with the weak-$*$ closure of $\hat{E}$. We can now state the Herglotz-type formula.

Theorem 2.1 *For each* $f \in \mathcal{M}$, *there exists a unique* $\mu = \mu_f \in P_E(\mathbb{S})$ *such that*

$$f(z) = \int_{\mathbb{S}} \left(\frac{2n(u-z)}{(1 - \langle z, u \rangle)^{n+1}} - 2nu - z \right) d\mu(u), \qquad z \in \mathbb{B}. \tag{2.1}$$

The mapping $f \mapsto \mu_f$ *is an affine homeomorphism of* $\mathcal{M}$ *onto its range when* $\mathcal{M}$ *and* $P_E(\mathbb{S})$ *are respectively endowed with the topology of uniform convergence on compact subsets of* $\mathbb{B}$ *and the weak-$*$ topology.*

Remark 2.2 When $n = 1$, the monomials spanned to obtain E are simply $u \mapsto u^m$ for $u \in \partial\mathbb{D}$ and $m \in \mathbb{Z}$. That is, E consists of the trigonometric polynomials.

Density of the trigonometric polynomials in $C(\partial\mathbb{D})$ and the Hahn–Banach theorem give $P_E(\partial\mathbb{D}) = P(\partial\mathbb{D})$. The above formula reduces to

$$f(z) = \int_{\partial\mathbb{D}} \frac{z(u+z)}{u-z} \, d\mu(u)$$

for $\mu \in P(\partial\mathbb{D})$, which is (1.1) re-expressed for $\mathcal{M}$ instead of $\mathcal{P}$.

Remark 2.3 The insistence that $\mu \in P_E(\mathbb{S})$ instead of simply requiring $\mu \in P(\mathbb{S})$ is done to guarantee uniqueness in the representation when $n \geq 2$. For example, let $\varphi(u) = u_1^2 \bar{u}_2^2$ for $u \in \mathbb{S}$ and $\omega \in M(\mathbb{S})$ be the signed measure given by $d\omega = \operatorname{Re} \varphi \, d\sigma$. Because holomorphic monomials restricted to $\mathbb{S}$ are orthogonal in $L^2(\sigma)$, $\omega(\mathbb{S}) = 0$. Since $|\omega(A)| \leq \sigma(A)$ for any Borel set $A \subseteq \mathbb{S}$, we get that $\mu = \sigma + \omega \in P(\mathbb{S})$. One can show that both σ and μ represent the identity mapping $f = I$ in (2.1).

The motivation behind the proof of Theorem 2.1 is that the one-variable result, Theorem 1.1, follows from the Poisson formula

$$g(z) = \frac{1}{2\pi} \int_0^{2\pi} \frac{e^{it}+z}{e^{it}-z} \operatorname{Re} g(e^{it}) \, dt, \qquad z \in \mathbb{D}, \tag{2.2}$$

valid for $g \in H(\mathbb{D}, \mathbb{C}) \cap C(\overline{\mathbb{D}})$ such that $g(0) \in \mathbb{R}$. Let $A(\mathbb{B}, \mathbb{C}^n)$ be the space consisting of those continuous $f : \overline{\mathbb{B}} \to \mathbb{C}^n$ for which $f|_{\mathbb{B}} \in H(\mathbb{B}, \mathbb{C}^n)$. We first developed the Cauchy-type formula

$$f(z) = \int_{\mathbb{S}} \frac{n(u-z)}{(1-\langle z,u\rangle)^{n+1}} \langle f(u), u \rangle \, d\sigma(u), \qquad z \in \mathbb{B},$$

for $f \in A(\mathbb{B}, \mathbb{C}^n)$, which reproduces f by way of a fixed vector-valued kernel and the scalar function $u \mapsto \langle f(u), u \rangle$, $u \in \mathbb{S}$, whose range is the numerical range of f. From this, we then obtained an analog of (2.2),

$$f(z) = \int_{\mathbb{S}} \left(\frac{2n(u-z)}{(1-\langle z,u\rangle)^{n+1}} - 2nu - z \right) \operatorname{Re}\langle f(u), u \rangle \, d\sigma(u), \qquad z \in \mathbb{B},$$

for all $f \in A(\mathbb{B}, \mathbb{C}^n)$ such that $f(0) = 0$ and $Df(0) = \lambda I$ for some $\lambda \in \mathbb{R}$. A standard limiting argument gives (2.1) for some $\mu \in P(\mathbb{S})$, and special care was taken to ensure that $\mu \in P_E(\mathbb{S})$. Uniqueness was then argued from an analysis of the measures in $P_E(\mathbb{S})$.

Let $P_H(\mathbb{S}) \subseteq P_E(\mathbb{S})$ denote the set of measures obtained through the representation in Theorem 2.1. Since the class $\mathcal{M}$ is convex and compact in the topology of uniform convergence on compact subsets of $\mathbb{B}$, $P_H(\mathbb{S})$ is convex and compact in the weak-$*$ topology.

Remark 2.4 It can be shown through a direct calculation that any $\mu \in P(\mathbb{S})$ such that (2.1) holds for some $f \in \mathcal{M}$ satisfies the orthogonality condition

$$\int_{\mathbb{S}} u_j \bar{u}_k \, d\mu(u) = \begin{cases} 1/n & \text{if } j = k, \\ 0 & \text{if } j \neq k \end{cases}$$

for $j, k \in \{1, \ldots, n\}$.

Suppose that $\mu \in M(\mathbb{S})$ is given by $d\mu(u) = n|u_1|^2 \, d\sigma(u)$. It follows from standard identities (see [6, 8] or (2.3) below) that $\mu(\mathbb{S}) = 1$. Since $u \mapsto |u_1|^2$ is in E, we have $\mu \in P_E(\mathbb{S})$. However,

$$\int_{\mathbb{S}} |u_1|^2 \, d\mu(u) = n \int_{\mathbb{S}} |u_1|^4 \, d\sigma(u) = \frac{2}{n+1} \neq \frac{1}{n}$$

if $n \geq 2$. This shows that $P_H(\mathbb{S})$ is a proper subset of $P_E(\mathbb{S})$ for such n.

We now can state our first open question.

Open Question 2.5 Can we characterize the measures in $P_H(\mathbb{S})$ in a useful way?

In the spirit of this question, we observe the following characteristics of measures in $P_H(\mathbb{S})$ that are already known. (See [4].)

- If $\mu \in P_E(\mathbb{S})$ and $\alpha, \beta \in \mathbb{N}_0^n$, then

$$\int_{\mathbb{S}} u^\alpha \bar{u}^\beta \, d\mu(u) = 0$$

  if there are $j, k \in \{1, \ldots, n\}$ such that $\alpha_j - \beta_j \geq 2$ and $\beta_k - \alpha_k \geq 2$.
- If $\mu \in P_H(\mathbb{S})$ and $\alpha \in \mathbb{N}_0^n$, then

$$\int_{\mathbb{S}} |u^\alpha|^2 u_j \bar{u}_k \, d\mu(u) = 0$$

  whenever $j, k \in \{1, \ldots, n\}$ are distinct.
- For all $\mu \in P_H(\mathbb{S})$ and $\alpha \in \mathbb{N}_0^n$, we have

$$\int_{\mathbb{S}} |u^\alpha|^2 \, d\mu(u) = \frac{(n-1)!\alpha!}{(n+|\alpha|-1)!}, \tag{2.3}$$

  where $\alpha! = \prod_{k=1}^n \alpha_k!$ and $|\alpha| = \sum_{k=1}^n \alpha_k$. (This is a standard identity for $\mu = \sigma$. See [6, 8].)
- If $n \geq 2$ and $C_v = \{e^{it}v : 0 \leq t \leq 2\pi\}$ for some $v \in \mathbb{S}$, then $\mu(C_v) = 0$ for all $\mu \in P_H(\mathbb{S})$. In particular, $P_H(\mathbb{S})$ contains no Dirac measures when $n \geq 2$. (Recall that the Dirac measures are the extreme points of $P(\partial\mathbb{D}) = P_H(\partial\mathbb{D})$ when $n = 1$.)

We conclude the section with more open questions.

Open Question 2.6 The set C_v given above is the intersection of a one-complex-dimensional subspace of $\mathbb{C}^n$ with $\mathbb{S}$. What is the highest dimension (complex or real) of a subspace of $\mathbb{C}^n$ whose intersection A with $\mathbb{S}$ has $\mu(A) = 0$ for all $\mu \in P_H(\mathbb{S})$?

Open Question 2.7 If we can characterize $P_H(\mathbb{S})$, can we use the characterization to find the extreme points of $P_H(\mathbb{S})$? This would immediately provide a representation of the extreme points of the class $\mathcal{M}$.

Open Question 2.8 Are there any as-of-now unknown properties of $\mathcal{M}$ that can be found through the representation (2.1) even in the weaker case that μ is assumed only to be in $P_E(\mathbb{S})$, or even just in $P(\mathbb{S})$, instead of $P_H(\mathbb{S})$?

3 Application to Convex Mappings

Here, we consider the family of normalized convex mappings on the unit ball $\mathbb{B}$, that is, the class $\mathcal{K}(\mathbb{B})$ of those $f \in H(\mathbb{B}, \mathbb{C}^n)$ that are biholomorphic, satisfy $f(0) = 0$ and $Df(0) = I$, and are such that $f(\mathbb{B})$ is a convex domain in $\mathbb{C}^n$.

When $n = 1$, the class of convex functions $\mathcal{K}(\mathbb{D})$ has a deep and elegant theory, a significant piece of which was provided by Brickman et al. [1] when they used the Herglotz representation to express functions in $\mathcal{K}(\mathbb{D})$ as integrals of a fixed kernel with respect to some $\mu \in P(\partial\mathbb{D})$. We begin by reviewing their general approach while adding a piece of notation and an observation that will help us when moving to the higher-dimensional setting.

Define the set

$$\mathcal{R}(\mathbb{D}) = \left\{ f \in H(\mathbb{D}, \mathbb{C}) : f(0) = 0, \ f'(0) = 1, \text{ and } \operatorname{Re} \frac{f(z)}{z} > \frac{1}{2} \text{ if } z \in \mathbb{D} \right\}.$$

Due to a classical result of Marx and Strohhäcker [3, 7], $\mathcal{K}(\mathbb{D}) \subseteq \mathcal{R}(\mathbb{D})$. In [1], the authors used the Herglotz formula to represent functions $f \in \mathcal{R}(\mathbb{D})$ as

$$f(z) = \int_{\partial\mathbb{D}} \frac{z}{1 - \bar{u}z} \, d\mu(u), \qquad z \in \mathbb{D},$$

for $\mu \in P(\partial\mathbb{D})$. Through this correspondence, the extreme points of the convex set $\mathcal{R}(\mathbb{D})$ are produced by the Dirac measures δ_α for $\alpha \in \partial\mathbb{D}$, and hence are seen to be

$$z \mapsto \frac{z}{1 - \bar{\alpha}z}, \qquad z \in \mathbb{D}.$$

These extreme points are half-plane mappings and hence lie in $\mathcal{K}(\mathbb{D})$. It follows from the Krein–Milman theorem (noting that $H(\mathbb{D}, \mathbb{C})$ is a locally convex topological vector space) that $\overline{\operatorname{co}} \, \mathcal{K}(\mathbb{D}) = \mathcal{R}(\mathbb{D})$, where $\overline{\operatorname{co}} \, \mathcal{K}(\mathbb{D})$ denotes the closed convex hull of $\mathcal{K}(\mathbb{D})$ and hence the extreme points of $\overline{\text{co}}\,\mathcal{K}(\mathbb{D})$ are the above half-plane mappings.

For $n \geq 2$, we generalize $\mathcal{R}(\mathbb{D})$ by defining

$$\mathcal{R}(\mathbb{B}) = \bigg\{ f \in H(\mathbb{B}, \mathbb{C}^n) : f(0) = 0,\ Df(0) = I,$$

$$\text{and } \operatorname{Re}\langle f(z), z \rangle > \frac{\|z\|^2}{2} \text{ if } z \in \mathbb{B} \setminus \{0\} \bigg\}.$$

Below, we will give a proof of an unpublished result of Ted J. Suffridge that $\mathcal{K}(\mathbb{B}) \subseteq \mathcal{R}(\mathbb{B})$.

Key to the upcoming proof is the family of quasi-convex mappings of type A, which properly contains $\mathcal{K}(\mathbb{B})$ when $n \geq 2$. For a locally biholomorphic $f \in H(\mathbb{B}, \mathbb{C}^n)$ with $f(0) = 0$ and $Df(0) = I$ and for $u \in \mathbb{S}$, we define the function $G_{f,u} : \mathbb{D} \times \mathbb{D} \to \mathbb{C}_\infty$ by

$$G_{f,u}(\alpha, \beta) = \frac{2\alpha}{\langle Df(\alpha u)^{-1}[f(\alpha u) - f(\beta u)], u \rangle} - \frac{\alpha + \beta}{\alpha - \beta},$$

where $\mathbb{C}_\infty$ refers to the extended complex plane. The quasi-convex family is

$$\mathcal{G}(\mathbb{B}) = \{ f \in H(\mathbb{B}, \mathbb{C}^n) : f \text{ is locally biholomorphic}, f(0) = 0,\ Df(0) = I,$$

$$\text{and } \operatorname{Re} G_{f,u}(\alpha, \beta) > 0 \text{ if } u \in \mathbb{S} \text{ and } \alpha, \beta \in \mathbb{D} \}.$$

We take note that this definition differs slightly from the original definition of Roper and Suffridge [5] and from the definition given in [2]. In [5], the family $\mathcal{G}(\mathbb{B})$ is defined to consist of biholomorphic mappings instead of locally biholomorphic mappings. It that article, Roper and Suffridge proved that if $f \in \mathcal{G}(\mathbb{B})$, then f is a starlike (biholomorphic) mapping, and, as noted in [2], f is actually starlike of order $1/2$, a generalization of another classical one-variable result of Marx and Strohhäcker [3, 7]. Since the proof of that result only requires the assumption of a locally biholomorphic f, we use that more general notion in our definition. In [2], the family $\mathcal{G}(\mathbb{B})$ is defined to consist of those normalized locally biholomorphic f such that $\operatorname{Re} G_{f,u}(\alpha, \beta) > 0$ for all $u \in \mathbb{S}$ but only for $\alpha, \beta \in \mathbb{D}$ satisfying $|\beta| < |\alpha|$. While that condition is nicely compatible with the conditions used to prove the above starlikeness result and the containment of $\mathcal{K}(\mathbb{B})$ in $\mathcal{G}(\mathbb{B})$, we will make use of the more restrictive condition given in our definition, which is consistent with that given in [5], in the upcoming proof.

Since $\mathcal{K}(\mathbb{B}) \subseteq \mathcal{G}(\mathbb{B})$ (see [2, 5]), the following result gives the desired containment. Although our focus is on the Euclidean ball, this argument easily generalizes to unit balls in alternative norms.

Theorem 3.1 *We have* $\mathcal{G}(\mathbb{B}) \subseteq \mathcal{R}(\mathbb{B})$.

Proof Let $f \in \mathcal{G}(\mathbb{B})$ and $u \in \mathbb{S}$. It suffices to verify the inequality $\mathrm{Re}\langle f(z), z\rangle > \|z\|^2/2$ for $z = \beta u$, $\beta \in \mathbb{D} \setminus \{0\}$. The function $\mathbb{D} \ni \beta \mapsto \langle f(\beta u), u\rangle = \beta + O(\beta^2)$ (the last form following from the normalization of f) has an isolated zero at $\beta = 0$ and hence has isolated zeros in $\mathbb{D}$.

Fix $\beta \in \mathbb{D} \setminus \{0\}$ such that $\langle f(\beta u), u\rangle \neq 0$, and let $\varphi = G_{f,u}(\cdot, \beta)$. Since $G_{f,u}$ is holomorphic in the polydisk $\mathbb{D} \times \mathbb{D}$ [5], $\varphi \in H(\mathbb{D}, \mathbb{C})$. By hypothesis, $\mathrm{Re}\,\varphi(\alpha) > 0$ for all $\alpha \in \mathbb{D}$, and the normalization of f gives $\varphi(0) = 1$. Therefore $\varphi \in \mathcal{P}$.

It is well known (and can easily be proved using Theorem 1.1) that $|\varphi'(0)| \leq 2$. Therefore

$$2 \geq \lim_{\alpha \to 0} \left| \frac{\varphi(\alpha) - 1}{\alpha} \right|$$

$$= \lim_{\alpha \to 0} \frac{1}{|\alpha|} \left| \frac{2\alpha}{\langle Df(\alpha u)^{-1}[f(\alpha u) - f(\beta u)], u\rangle} - \frac{2\alpha}{\alpha - \beta} \right|$$

$$= 2 \left| \frac{1}{\beta} - \frac{1}{\langle f(\beta u), u\rangle} \right|.$$

We conclude that

$$\left| \frac{\beta}{\langle f(\beta u), u\rangle} - 1 \right| \leq |\beta| < 1 \tag{3.1}$$

for all $\beta \in \mathbb{D} \setminus \{0\}$ such that $\langle f(\beta u), u\rangle \neq 0$. Were there an isolated zero of $\beta \mapsto \langle f(\beta u), u\rangle$ in $\mathbb{D} \setminus \{0\}$, letting $\beta \in \mathbb{D} \setminus \{0\}$ tend to that zero would violate (3.1). Hence $\langle f(\beta u), u\rangle \neq 0$ for all $\beta \in \mathbb{D} \setminus \{0\}$, and (3.1) holds for all such β.

The reciprocal map takes the disk $\{\lambda \in \mathbb{C} : |\lambda - 1| < 1\}$ onto the half-plane $\{w \in \mathbb{C} : \mathrm{Re}\,w > 1/2\}$. Therefore

$$\frac{1}{2} < \mathrm{Re}\, \frac{\langle f(\beta u), u\rangle}{\beta} = \frac{\mathrm{Re}\langle f(\beta u), \beta u\rangle}{|\beta|^2}, \qquad \beta \in \mathbb{D} \setminus \{0\},$$

which is what we set out to prove. □

Now $f \in \mathcal{R}(\mathbb{B})$ if and only if $2f - I \in \mathcal{M}$, which gives

$$2f(z) - z = \int_{\mathbb{S}} \left(\frac{2n(u - z)}{(1 - \langle z, u\rangle)^{n+1}} - 2nu - z \right) d\mu(u), \qquad z \in \mathbb{B},$$

for some $\mu \in P_H(\mathbb{S})$. The compatibility of the two sides of the above equation is notable, and it allows us to immediately conclude the following.

Theorem 3.2 *Let $f\colon \mathbb{B} \to \mathbb{C}^n$. Then $f \in \mathcal{R}(\mathbb{B})$ if and only if there is a $\mu = \mu_f \in P_H(\mathbb{S})$ such that*

$$f(z) = \int_{\mathbb{S}} \left(\frac{n(u-z)}{(1 - \langle z, u \rangle)^{n+1}} - nu \right) d\mu(u), \qquad z \in \mathbb{B}. \tag{3.2}$$

The mapping $f \mapsto \mu_f$ is an affine homeomorphism of $\mathcal{R}(\mathbb{B})$ onto $P_H(\mathbb{S})$ when $\mathcal{R}(\mathbb{B})$ and $P_H(\mathbb{S})$ are respectively endowed with the topology of uniform convergence on compact subsets of $\mathbb{B}$ and the weak-$*$ topology.

Of course, this also provides a representation of mappings in $\mathcal{G}(\mathbb{B})$ and in $\mathcal{K}(\mathbb{B})$ using subsets of measures in $P_H(\mathbb{S})$. We can now ask the following questions.

Open Question 3.3 Clearly $\mathcal{R}(\mathbb{B})$ is convex and compact in the topology of uniform convergence on compact sets. Is $\mathcal{R}(\mathbb{B}) = \overline{\mathrm{co}}\,\mathcal{K}(\mathbb{B})$? Is $\mathcal{R}(\mathbb{B}) = \overline{\mathrm{co}}\,\mathcal{G}(\mathbb{B})$?

Open Question 3.4 If $\mathcal{R}(\mathbb{B}) \neq \overline{\mathrm{co}}\,\mathcal{K}(\mathbb{B})$, can we characterize the measures in $P_H(\mathbb{S})$ corresponding to mappings in $\mathcal{K}(\mathbb{B})$ or in $\overline{\mathrm{co}}\,\mathcal{K}(\mathbb{B})$ through (3.2)? The analogous question can be asked for $\mathcal{G}(\mathbb{B})$.

References

1. Brickman, L., MacGregor, T.H., Wilken, D.R.: Convex hulls of some classical families of univalent functions. Trans. Am. Math. Soc. **156**, 91–107 (1971)
2. Graham, I., Kohr, G.: Geometric Function Theory in One and Higher Dimensions. Marcel Dekker, New York (2003)
3. Marx, A.: Untersuchungen über schlichte Abbildungen. Math. Ann. **107**, 40–67 (1932)
4. Muir, Jr., J.R.: A Herglotz-type representation for vector-valued holomorphic mappings on the unit ball of $\mathbb{C}^n$. J. Math. Anal. Appl. **440**, 127–144 (2016)
5. Roper, K.A., Suffridge, T.J.: Convexity properties of holomorphic mappings in $\mathbb{C}^n$. Trans. Am. Math. Soc. **351**, 1803–1833 (1999)
6. Rudin, W.: Function Theory in the Unit Ball of $\mathbb{C}^n$. Springer, New York (1980)
7. Strohhäcker, E.: Beitrage zur Theorie der schlichten Funktionen. Math. Z. **37**, 356–380 (1933)
8. Zhu, K.: Spaces of Holomorphic Functions in the Unit Ball. Springer, New York (2005)

Extremal Problems and Convergence Results for Mappings with Generalized Parametric Representation in $\mathbb{C}^n$

Hidetaka Hamada, Mihai Iancu, and Gabriela Kohr

Abstract In this paper we survey recent results related to extremal problems for the family $\widetilde{S}_A^t(\mathbb{B}^n)$ of normalized univalent mappings on the Euclidean unit ball $\mathbb{B}^n$ in $\mathbb{C}^n$, which have generalized parametric representation with respect to time-dependent operators $A \in \widetilde{\mathcal{A}}$, where $\widetilde{\mathcal{A}}$ is the family of all measurable mappings $A : [0, \infty) \rightarrow L(\mathbb{C}^n)$, which satisfy certain natural conditions. In the second part of this paper, we consider the dependence of $\widetilde{S}_A^t(\mathbb{B}^n)$ on $t \geq 0$ and on $A \in \widetilde{\mathcal{A}}$, and we present some convergence results related to the family $\widetilde{S}_A^t(\mathbb{B}^n)$ in terms of the Carathéodory metric ρ on $H(\mathbb{B}^n)$. Various questions and remarks are also provided, which point out main differences between the usual parametric representation with respect to time-independent operators and that with respect to time-dependent operators.

Keywords Carathéodory family · Extreme point · Generalized parametric representation · Loewner chain · Loewner differential equation · Support point

2000 Mathematics Subject Classification Primary 32H02; Secondary 30C45

1 Introduction and Preliminaries

Let $\mathbb{C}^n$ be the space of n complex variables $z = (z_1, \ldots, z_n)$ with the Euclidean inner product $\langle z, w \rangle = \sum_{j=1}^{n} z_j \overline{w}_j$ and the Euclidean norm $\|z\| = \langle z, z \rangle^{1/2}$. The open unit ball $\{z \in \mathbb{C}^n : \|z\| < 1\}$ is denoted by $\mathbb{B}^n$. In the case $n = 1$, the unit disc $\mathbb{B}^1$ is denoted by $\mathbb{U}$.

H. Hamada
Faculty of Science and Engineering, Kyushu Sangyo University, Higashi-ku Fukuoka, Japan
e-mail: h.hamada@ip.kyusan-u.ac.jp

M. Iancu · G. Kohr (✉)
Faculty of Mathematics and Computer Science, Babeș-Bolyai University, Cluj-Napoca, Romania
e-mail: miancu@math.ubbcluj.ro; gkohr@math.ubbcluj.ro

© Springer International Publishing AG, part of Springer Nature 2017
F. Bracci (ed.), *Geometric Function Theory in Higher Dimension*,
Springer INdAM Series 26, https://doi.org/10.1007/978-3-319-73126-1_9

Let $L(\mathbb{C}^n)$ denote the space of linear operators from $\mathbb{C}^n$ into $\mathbb{C}^n$ with the standard operator norm. Also, let I_n be the identity operator in $L(\mathbb{C}^n)$. If $A \in L(\mathbb{C}^n)$, we denote by A^* the adjoint of the operator A.

Let $H(\mathbb{B}^n)$ be the family of holomorphic mappings from $\mathbb{B}^n$ into $\mathbb{C}^n$ with the compact-open topology. If $f \in H(\mathbb{B}^n)$, we say that f is normalized if $f(0) = 0$ and $Df(0) = I_n$. Let $S(\mathbb{B}^n)$ be the family of normalized biholomorphic mappings on $\mathbb{B}^n$. If $n = 1$, then the family $S(\mathbb{U})$ is denoted by S.

We shall use the following notations related to an operator $A \in L(\mathbb{C}^n)$ (cf. [7]):

$$m(A) = \min\{\Re\langle A(z), z\rangle : \|z\| = 1\},$$

$$k_+(A) = \max\{\Re\lambda : \lambda \in \sigma(A)\},$$

where $\sigma(A)$ is the spectrum of A and $k_+(A)$ is the Lyapunov index of A. Then $m(A) \le k_+(A) \le \|A\|$ (see e.g. [7]).

Let $A \in L(\mathbb{C}^n)$ be such that $m(A) \ge 0$. The following family of holomorphic mappings on $\mathbb{B}^n$ plays the role of the Carathéodory family in $\mathbb{C}^n$ (see [19, 26]):

$$\mathcal{N}_A = \{h \in H(\mathbb{B}^n) : h(0) = 0, Dh(0) = A, \Re\langle h(z), z\rangle \ge 0, z \in \mathbb{B}^n\}.$$

Various applications concerning the family $\mathcal{N}_A$ in the theory of Loewner chains and univalence in $\mathbb{C}^n$ may be found in [1, 3, 6, 10, 11, 13–16, 19, 22, 23, 25, 28].

Let $A \in L(\mathbb{C}^n)$ be such that $m(A) > 0$. We denote by $\widehat{S}_A(\mathbb{B}^n)$ the family of A-spirallike mappings on $\mathbb{B}^n$ (see e.g. [22] and [26]):

$$\widehat{S}_A(\mathbb{B}^n) = \{f \in S(\mathbb{B}^n) : e^{-tA}f(\mathbb{B}^n) \subseteq f(\mathbb{B}^n), t \ge 0\}.$$

Definition 1.1 Let $A : [0, \infty) \to L(\mathbb{C}^n)$ be a measurable mapping that is locally integrable on $[0, \infty)$. For every $s \ge 0$, we denote by $V(s, \cdot) : [s, \infty) \to L(\mathbb{C}^n)$ the unique locally absolutely continuous solution of the initial value problem (see [5]; cf. [27])

$$\frac{\partial V}{\partial t}(s, t) = -A(t)V(s, t), \quad \text{a.e. } t \in [s, \infty), \quad V(s, s) = I_n. \tag{1.1}$$

Also, let $V(t) = V(0, t)$ for $t \ge 0$. Then $V(s, t) = V(t)V(s)^{-1}$, for $0 \le s \le t < \infty$.

Remark 1.2 Let $A : [0, \infty) \to L(\mathbb{C}^n)$ be a measurable mapping which is locally integrable on $[0, \infty)$, and let $s \ge 0$. If $A(t)$ and $\int_s^t A(\tau)d\tau$ commute, for all $t \ge s$, then

$$V(s, t) = e^{-\int_s^t A(\tau)d\tau}, \quad \forall t \in [s, \infty), \tag{1.2}$$

in view of [5, Exercise VII.2.22] (cf. [11, Remark 1.6]). Therefore, if $A(t) = \mathbf{A} \in L(\mathbb{C}^n)$, for all $t \ge 0$, then $V(t) = e^{-t\mathbf{A}}$, $t \ge 0$.

Definition 1.3 ([15, 16]) Let $\widetilde{\mathcal{A}}$ be the family of all measurable mappings A : $[0, \infty) \to L(\mathbb{C}^n)$ which satisfy the following conditions:

  (i) $m(A(\tau)) \geq 0$, for a.e. $\tau \geq 0$;
 (ii) $\operatorname{ess\,sup}_{s \geq 0} \|A(s)\| < \infty$;
(iii) $\sup_{s \geq 0} \int_s^\infty \|V(s,t)^{-1}\| e^{-2\int_s^t m(A(\tau))d\tau} dt < \infty$, where $V(s,t)$ is the unique solution on $[s, \infty)$ of the initial value problem (1.1) related to A.

Remark 1.4 In the case that $A(t) \equiv \mathbf{A} \in L(\mathbb{C}^n)$ is a time-independent operator, it is easily seen that $A \in \widetilde{\mathcal{A}}$ if and only if $k_+(\mathbf{A}) < 2m(\mathbf{A})$. This condition was successfully used by the authors in [10, 13, 14], to obtain various properties of mappings with **A**-parametric representation on $\mathbb{B}^n$.

The set of conditions in Definition 1.3 is suitable for the study of the mappings with generalized parametric representation and the normal Loewner chains with respect to time-dependent operators, as we shall see in the following section.

Note that an example of a time-dependent operator in $\widetilde{\mathcal{A}}$ which does not satisfy (1.2) may be found in [16, Example 2.11].

The following remark is useful in providing a necessary and sufficient condition for a certain time-dependent operator to belong to the family $\widetilde{\mathcal{A}}$ (see [16]). This result yields concrete examples of operators $A \in \widetilde{\mathcal{A}}$.

Remark 1.5 Let $T > 0$, $\mathbf{A} \in L(\mathbb{C}^n)$, and let $A : [0, \infty) \to L(\mathbb{C}^n)$ be a measurable mapping such that $m(A(t)) \geq 0$, for a.e. $t \in [0, T]$, $\operatorname{ess\,sup}_{t \in [0,T]} \|A(t)\| < \infty$, and $A(t) = \mathbf{A}$, for a.e. $t > T$. Then $A \in \widetilde{\mathcal{A}}$ if and only if $k_+(\mathbf{A}) < 2m(\mathbf{A})$.

2 Loewner Chains and Generalized Parametric Representation

In this section we consider the definition of a generalized parametric representation with respect to a time-dependent linear operator, in the case of normalized holomorphic mappings on $\mathbb{B}^n$. This notion was first considered by Graham, Hamada, Kohr and Kohr [11], and then generalized by Hamada, Iancu and Kohr [16] (cf. [27, Proposition 1.5.1]) (see [9] and [21], in the case of $A = I_n$).

Definition 2.1 Let $A \in \widetilde{\mathcal{A}}$ and let $V(s,t)$ be the unique solution on $[s, \infty)$ of the initial value problem (1.1) related to A. Let $T \geq 0$. We say that a mapping $f : \mathbb{B}^n \to \mathbb{C}^n$ has generalized parametric representation with respect to A on $[T, \infty)$ if there exists a mapping $h = h(z, t) : \mathbb{B}^n \times [0, \infty) \to \mathbb{C}^n$, called a Herglotz vector field with respect to A (cf. [6]), with

  (i) $h(z, \cdot)$ is measurable on $[0, \infty)$, for all $z \in \mathbb{B}^n$,
 (ii) $h(\cdot, t) \in \mathcal{N}_{A(t)}$, for all $t \geq 0$,

such that

$$f(z) = \lim_{t \to \infty} V(T, t)^{-1} v(z, T, t)$$

locally uniformly on $\mathbb{B}^n$, where $v(z, T, \cdot) : [T, \infty) \to \mathbb{C}^n$ is the unique locally absolutely continuous solution of the initial value problem

$$\frac{\partial v}{\partial t}(z, T, t) = -h(v(z, T, t), t), \quad \text{a.e. } t \in [T, \infty), \ v(z, T, T) = z, \tag{2.1}$$

for all $z \in \mathbb{B}^n$.

Let $\widetilde{S}_A^T(\mathbb{B}^n)$ be the family of mappings with generalized parametric representation with respect to A on $[T, \infty)$.

Remark 2.2 Let $\mathbf{A} \in L(\mathbb{C}^n)$ be such that $k_+(\mathbf{A}) < 2m(\mathbf{A})$ and let $A : [0, \infty) \to L(\mathbb{C}^n)$ be such that $A(t) = \mathbf{A}$, for all $t \geq 0$. Then $A \in \widetilde{\mathcal{A}}$ and $\widetilde{S}_A^T(\mathbb{B}^n)$ reduces to the family $S_{\mathbf{A}}^0(\mathbb{B}^n)$ of mappings with $\mathbf{A}$-parametric representation on $\mathbb{B}^n$, for all $T \geq 0$ (see [10, 13]). In particular, if $\mathbf{A} \equiv I_n$, then $\widetilde{S}_{I_n}^T(\mathbb{B}^n) = S^0(\mathbb{B}^n)$, for all $T \geq 0$, where $S^0(\mathbb{B}^n)$ is the family of mappings with the usual parametric representation on $\mathbb{B}^n$ (see [8]). This family has been deeply studied and various unexpected properties regarding $S^0(\mathbb{B}^n)$ have been developed (see [2, 4, 8, 9, 12]).

In the case $n = 1, S^0(\mathbb{U}) = S$, but if $n \geq 2$, then $S^0(\mathbb{B}^n)$ is a proper subset of $S(\mathbb{B}^n)$ (see [9]). However, important properties of the family S related to growth, covering, compactness, and embedding results in Loewner chains, also hold in the case of the family $S^0(\mathbb{B}^n)$, $n \geq 2$ (see e.g. [8, Chapter 8]). On the other hand, significant differences between the one variable theory and that in higher dimensions have been recently obtained (see [2–4, 6, 9, 12, 16, 23, 28]).

In the following, we consider the definition of a normal Loewner chain with respect to a time-dependent operator (see [16]; cf. [11, 27]).

Definition 2.3 Let $A \in \widetilde{\mathcal{A}}$, let $V(t) = V(0, t)$ be the unique solution on $[0, \infty)$ of the initial value problem (1.1) related to A, and let $f : \mathbb{B}^n \times [0, \infty) \to \mathbb{C}^n$ be a mapping. We say that f is a normal Loewner chain with respect to A, if $f(\mathbb{B}^n, s) \subseteq f(\mathbb{B}^n, t)$, for $0 \leq s \leq t, f(\cdot, t)$ is univalent on $\mathbb{B}^n$ with $f(0, t) = 0$, $Df(0, t) = V(t)^{-1}$, for $t \geq 0$, and $\{V(t)f(\cdot, t)\}_{t \geq 0}$ is a normal family on $\mathbb{B}^n$.

If, in addition, $A(t) = \mathbf{A} \in L(\mathbb{C}^n), t \geq 0$, is a time-independent operator such that $k_+(\mathbf{A}) < 2m(\mathbf{A})$, then we say that f is an $\mathbf{A}$-normal Loewner chain, and if $\mathbf{A} = I_n$, then f is a usual normal Loewner chain.

Remark 2.4 We remark that in the case of one complex variable, every non-normalized Loewner chain may be easily transformed into a normalized one by a suitable change of variable, and the new chain essentially preserves the geometric properties of the initial Loewner chain (see [20, Chapter 6]). However, in dimension $n \geq 2$, there exist non-normalized Loewner chains (even with respect to a time-independent operator), which cannot be normalized in the same simple way as in the case of one complex variable (see [10]).

Loewner chains and the associated Loewner differential equations in higher dimensions were first studied by Pfaltzgraff [19].

The connection between the notions of generalized parametric representation and normal Loewner chains with respect to time-dependent operators in $\widetilde{\mathcal{A}}$ was obtained in [11] and [27] (see also [16]). The particular case of a time-independent operator $A \in L(\mathbb{C}^n)$ with $k_+(A) < 2m(A)$ was considered in [10] (cf. [9], [8, Section 8.1.1]). We mention that Theorem 2.5 below (see [16]) does not hold in general, for time-dependent operators which are not in $\widetilde{\mathcal{A}}$, not even for time-independent operators (see [28, Example 3.7]).

Theorem 2.5 *Let $A \in \widetilde{\mathcal{A}}$ and let $V(s, t)$ be the unique solution on $[s, \infty)$ of the initial value problem (1.1) related to A. Let h be a Herglotz vector field with respect to A and let $v(z, T, \cdot) : [T, \infty) \to \mathbb{C}^n$ be the unique locally absolutely continuous solution of the initial value problem (2.1) associated to h. Then the limit*

$$f(z, T) = \lim_{t \to \infty} V(T, t)^{-1} v(z, T, t)$$

exists locally uniformly with respect to $z \in \mathbb{B}^n$, for all $T \geq 0$. Moreover, f is a normal Loewner chain with respect to A.

The following theorem provides a characterization of the family of mappings with generalized parametric representation with respect to a time-dependent operator and for a given time (see [16]). The given time plays a crucial role in this result for time-dependent operators in $\widetilde{\mathcal{A}}$, which is in contrast to the case of time-independent operators $A \in L(\mathbb{C}^n)$ with $k_+(A) < 2m(A)$ (see [10, 13]; cf. Remark 2.2), as we shall point out later on in this section.

Theorem 2.6 *Let $T \geq 0$, $A \in \widetilde{\mathcal{A}}$, and let $V(t) = V(0, t)$ and $V(s, t)$ be the unique locally absolutely continuous solution on $[s, \infty)$ of the initial value problem (1.1) related to A. Let $g \in H(\mathbb{B}^n)$ be a normalized mapping. Then $g \in \widetilde{S}_A^T(\mathbb{B}^n)$ if and only if there exists a normal Loewner chain f with respect to A such that $g = V(T)f(\cdot, T)$.*

The next theorem provides the compactness of the families of mappings with generalized parametric representation with respect to a time-dependent operator (see [16]). If $A(t) = \mathbf{A} \in L(\mathbb{C}^n)$, $t \geq 0$, is a time-independent operator such that $k_+(\mathbf{A}) < 2m(\mathbf{A})$ (in particular, if $A(t) = I_n$), see [10] and [8].

Theorem 2.7 *Let $T \geq 0$ and $A \in \widetilde{\mathcal{A}}$. Then $\widetilde{S}_A^T(\mathbb{B}^n)$ is a compact subset of $H(\mathbb{B}^n)$.*

We remark that the results contained in Theorems 2.5–2.7 justify the technical assumptions in Definition 1.3, which are optimal in the sense that they preserve some of the most important properties of the usual parametric representation and the normal Loewner chains (cf. [9]). However, some differences can be pointed out, and this leads to some new phenomena. The authors in [16] gave examples of nonconstant time-dependent operators $A \in \widetilde{\mathcal{A}}$ such that $\widetilde{S}_A^t(\mathbb{B}^2) \neq \widetilde{S}_A^s(\mathbb{B}^2)$, for some $s, t \in [0, \infty)$, $s \neq t$. In the following we give one of these examples (see [16, Examples 3.5]).

Example 2.8 Let $T > 0$, $\varepsilon \in (0,1)$, and $\mathbf{A} = \begin{pmatrix} 2 & 0 \\ 0 & 1+\varepsilon \end{pmatrix}$. Also, let $A \in \widetilde{\mathcal{A}}$ be

given by $A(t) = \begin{cases} \mathbf{A}, & t \in [0,T), \\ I_2, & t \in [T,\infty). \end{cases}$ Then $\widetilde{S}_A^t(\mathbb{B}^2) = S^0(\mathbb{B}^2)$, for all $t \in [T,\infty)$, and
$\widetilde{S}_A^0(\mathbb{B}^2) \neq S^0(\mathbb{B}^2)$, for sufficiently small $\varepsilon \in (0,1)$ and sufficiently large $T > 0$.

The following question, which is natural in view of Remark 2.2, is open (see [16]).

Question 2.9 *Let $A \in \widetilde{\mathcal{A}}$ and $T \geq 0$. Does there exist $\mathbf{A} \in L(\mathbb{C}^n)$ such that $k_+(\mathbf{A}) < 2m(\mathbf{A})$ and $\widetilde{S}_A^T(\mathbb{B}^n) = S_A^0(\mathbb{B}^n)$, for $n \geq 2$?*

Remark 2.10 (see [16]; cf. [8]) For $n = 1$ and $a \in \widetilde{\mathcal{A}}$, we have $\widetilde{S}_a^t(\mathbb{U}) = S$, for all $t \geq 0$.

The next result, which is related to the Question 2.9, was obtained in [16].

Proposition 2.11 *Let $a : [0,\infty) \to \mathbb{R}$ be a measurable function such that*

$$ess\,inf_{t \geq 0} a(t) > 0 \quad and \quad ess\,sup_{t \geq 0} a(t) < \infty.$$

Let $\mathbf{A} \in L(\mathbb{C}^n)$ be such that $k_+(\mathbf{A}) < 2m(\mathbf{A})$, and let $A : [0,\infty) \to L(\mathbb{C}^n)$ be given by $A(t) = a(t)\mathbf{A}$, for a.e. $t \geq 0$. Then $A \in \widetilde{\mathcal{A}}$ and $\widetilde{S}_A^T(\mathbb{B}^n) = S_A^0(\mathbb{B}^n)$, for all $T \geq 0$.

Taking into account the above proposition (see also [15, Example 4.2]), it is natural to ask the following question.

Question 2.12 *Let $n \geq 2$. Under which conditions related to $A \in \widetilde{\mathcal{A}}$, the following equality $\widetilde{S}_A^T(\mathbb{B}^n) = \widetilde{S}_A^0(\mathbb{B}^n)$ holds for all $T \geq 0$?*

The next result related to Question 2.12, was obtained in [15, Proposition 4.3], by using certain convergence results, which will be mentioned in a forthcoming section. In particular, if $A(t) = \mathbf{A} \in L(\mathbb{C}^n)$, $t \geq 0$, is a time-independent operator with $\mathbf{A} + \mathbf{A}^* = 2\alpha I_n$, for some $\alpha > 0$, then $S_{\mathbf{A}}^0(\mathbb{B}^n) = S^0(\mathbb{B}^n)$ (see [10]).

Proposition 2.13 *Let $\alpha : [0,\infty) \to \mathbb{R}$ be a measurable function such that*

$$ess\,inf_{t \geq 0} \alpha(t) > 0 \ and \ ess\,sup_{t \geq 0} \alpha(t) < \infty.$$

Let $A : [0,\infty) \to L(\mathbb{C}^n)$ be a measurable mapping such that $A(t) + A(t)^* = 2\alpha(t)I_n$, for a.e. $t \geq 0$. Then $A \in \widetilde{\mathcal{A}}$ and $\widetilde{S}_A^T(\mathbb{B}^n) = S^0(\mathbb{B}^n)$, for all $T \geq 0$.

In Proposition 2.13, we have $k_+(A(t)) < 2m(A(t))$ and $S_{A(t)}^0(\mathbb{B}^n) = S^0(\mathbb{B}^n)$, for a.e. $t \geq 0$ (see [10, Theorem 3.12]). Hence, we consider the following question (see [15]).

Question 2.14 *Let $\mathbf{A} \in L(\mathbb{C}^n)$ be such that $k_+(\mathbf{A}) < 2m(\mathbf{A})$. Also, let $A \in \widetilde{\mathcal{A}}$ be such that $k_+(A(t)) < 2m(A(t))$ and $S_{A(t)}^0(\mathbb{B}^n) = S_{\mathbf{A}}^0(\mathbb{B}^n)$, for a.e. $t \geq 0$. Is it true that $\widetilde{S}_A^T(\mathbb{B}^n) = S_{\mathbf{A}}^0(\mathbb{B}^n)$, for all $T \geq 0$ and $n \geq 2$?*

3 Extremal Problems for the Family $\widetilde{S}_A^t(\mathbb{B}^n)$

In this section, we consider certain questions and results related to extremal problems associated with the compact family $\widetilde{S}_A^s(\mathbb{B}^n)$, where $A \in \widetilde{\mathcal{A}}$ and $s \geq 0$. To this end, we recall the notions of extreme/support points associated with compact subsets of $H(\mathbb{B}^n)$ (see e.g. [5, 23]).

Definition 3.1 Let $E \subseteq H(\mathbb{B}^n)$ be a nonempty compact set.

(i) A point $f \in E$ is called an *extreme point* of E ($f \in \text{ex}\,E$) if $f = \lambda g + (1 - \lambda)h$, for some $\lambda \in (0, 1)$, $g, h \in E$, implies that $f = g = h$.

(ii) A point $f \in E$ is called a *support point* of E ($f \in \text{supp}\,E$) if there exists a continuous linear functional $L : H(\mathbb{B}^n) \to \mathbb{C}$ such that $\Re L|_E$ is nonconstant on E, and $\Re L(f) = \max\{\Re L(g) : g \in E\}$.

Remark 3.2 In the case $n = 1$, every extreme/support point of the compact family S is an unbounded single-slit mapping (see [20]). In higher dimensions, the situation is completely different, due to a recent result of Bracci [2]. His result yields that in dimension $n \geq 2$, there exist bounded support points of the compact family $S^0(\mathbb{B}^n)$.

Recent contributions related to extremal problems for univalent mappings on $\mathbb{B}^n$ ($n \geq 2$), which have parametric representation, may be found in [2, 3, 12, 14, 24].

The following result was obtained in [16], and is a generalization of [12, Theorem 2.1] and [13, Theorem 3.1] to the case of time-dependent operators (see [17] and [18], in the case $n = 1$).

Theorem 3.3 *Let $A \in \widetilde{\mathcal{A}}$ and $s \geq 0$. Let f be a normal Loewner chain with respect to A. If $V(s)f(\cdot, s) \in \text{ex}\,\widetilde{S}_A^s(\mathbb{B}^n)$, then $V(t)f(\cdot, t) \in \text{ex}\,\widetilde{S}_A^t(\mathbb{B}^n)$, for all $t \geq s$, where $V(t) = V(0, t)$ and $V(s, t)$ is the unique solution on $[s, \infty)$ of the initial value problem (1.1) related to A.*

The following result, due to Hamada, Iancu, and Kohr [16], is a generalization of [25, Theorem 1.1] and [13, Theorem 3.5] to the case of time-dependent operators (see [17] and [18], in the case $n = 1$).

Theorem 3.4 *Let $s \geq 0$ and $A \in \widetilde{\mathcal{A}}$ be such that $\text{ess inf}_{t \geq 0} m(A(t)) > 0$. Let f be a normal Loewner chain with respect to A. If $V(s)f(\cdot, s) \in \text{supp}\,\widetilde{S}_A^s(\mathbb{B}^n)$, then $V(t)f(\cdot, t) \in \text{supp}\,\widetilde{S}_A^t(\mathbb{B}^n)$, for all $t \geq s$, where $V(t) = V(0, t)$ and $V(s, t)$ is the unique solution on $[s, \infty)$ of the initial value problem (1.1) related to A.*

Question 3.5 *Does Theorem 3.4 hold for time-dependent operators $A \in \widetilde{\mathcal{A}}$ with $\text{ess inf}_{t \geq 0} m(A(t)) = 0$ and $n \geq 2$?*

Remark 3.6 Let $n \geq 2$. The role of the normalization of the normal Loewner chain in Theorem 3.4 was pointed out by an example of a normal Loewner chain F with respect to an operator $A \in \widetilde{\mathcal{A}}$ with $\text{ess inf}_{t \geq 0} m(A(t)) > 0$ such that the following conditions hold (see [16]):

(i) $V(t)F(\cdot, t) \in S^0(\mathbb{B}^n)$, for all $t \geq 0$,

(ii) $F(\cdot, 0) \in \operatorname{supp} S^0(\mathbb{B}^n)$, but $V(t)F(\cdot, t) \notin \operatorname{supp} S^0(\mathbb{B}^n)$, for all $t > 0$,

where $V(t) = V(0, t)$ is the unique solution on $[0, \infty)$ of the initial value problem (1.1) related to A.

Taking into account the above remark and [16, Examples 4.8 and 4.9], it would be interesting to give an answer to the following questions. In the case of time-independent operators, see [2–4].

Question 3.7 *Let $n \geq 2$. Does Remark 3.6 remain true in the case of extreme points?*

Question 3.8 *Let $n \geq 2$. Do there exist $A \in \widetilde{\mathcal{A}}$ and a normal Loewner chain F with respect to A such that $F(\cdot, 0) \notin \operatorname{supp} \widetilde{S}_A^0(\mathbb{B}^n)$ (resp. $F(\cdot, 0) \notin \operatorname{ex} \widetilde{S}_A^0(\mathbb{B}^n)$), but $V(t)F(\cdot, t) \in \operatorname{supp} \widetilde{S}_A^t(\mathbb{B}^n)$ (resp. $V(t)F(\cdot, t) \in \operatorname{ex} \widetilde{S}_A^t(\mathbb{B}^n)$), for all $t > 0$, where $V(t) = V(0, t)$ is the unique solution on $[0, \infty)$ of the initial value problem (1.1) related to A?*

The following example (see [16]) points out a normal Loewner chain with respect to a particular time-dependent operator $A \in \widetilde{\mathcal{A}}$, which generates bounded support points for the compact family $\widetilde{S}_A^s(\mathbb{B}^2)$. This result is a generalization of recent results in [2, Theorem 1.2] and [14, Theorem 7.6], and is in contrast to the one-dimensional case (see e.g. [4] and [12]).

Example 3.9 Let $T > 0$, $\lambda_1, \lambda_2 \in [1, 2)$, $\lambda : [0, \infty) \to [1, 2)$ be given by $\lambda(t) = \begin{cases} \lambda_1, & t \in [0, T) \\ \lambda_2, & t \in [T, \infty) \end{cases}$ and let $A(t) = \begin{pmatrix} \lambda(t) & 0 \\ 0 & 1 \end{pmatrix}$, for all $t \geq 0$. Also, let

$$\alpha(\lambda) = \frac{1}{2 - \lambda} \max \left\{ a > 0 : \lambda x^2 + y^2 - axy^2 \geq 0, x, y \geq 0, x^2 + y^2 \leq 1 \right\}, \ \lambda \in [1, 2),$$

and $\beta(s) = e^{(s-T)(2-\lambda_1)}$, $s \in [0, T]$.

Then $A \in \widetilde{\mathcal{A}}$, and the mapping $F : \mathbb{B}^2 \times [0, \infty) \to \mathbb{C}^2$ given by

$$V(s)F(z, s) = \begin{cases} \left(z_1 + \left(\alpha(\lambda_1)(1 - \beta(s)) + \alpha(\lambda_2)\beta(s) \right) z_2^2, z_2 \right), & 0 \leq s < T \\ \left(z_1 + \alpha(\lambda_2) z_2^2, z_2 \right), & s \geq T, \end{cases}$$

for all $z = (z_1, z_2) \in \mathbb{B}^2$, is a normal Loewner chain with respect to A, where $V(t) = V(0, t)$ is the unique solution on $[0, \infty)$ of the initial value problem (1.1) related to A. Moreover, $V(s)F(\cdot, s)$ is a bounded support point for $\widetilde{S}_A^s(\mathbb{B}^2)$, for all $s \geq 0$.

In view of Example 3.9, it is natural to ask the following question, which is still open even in the case of time-independent operators (see [3, 4, 14]).

Question 3.10 *Does the normal Loewner chain given in Example 3.9 generate also extreme points for the family $\widetilde{S}_A^s(\mathbb{B}^2)$, $s \geq 0$?*

4 Continuous Dependence of the Time-Dependent Normalization

We begin this section with the dependence of $\widetilde{S}_A^T(\mathbb{B}^n)$ on $T \geq 0$, when $A \in \widetilde{\mathcal{A}}$ is fixed (see [16]; cf. [13] and [23]). To this end, first we recall the definition of the Hausdorff metric on $H(\mathbb{B}^n)$ (cf. [23]).

Definition 4.1 Let δ be the well known metric on $H(\mathbb{B}^n)$ such that $(H(\mathbb{B}^n), \delta)$ is a Fréchet space with respect to the compact-open topology. For all nonempty subsets V and W of $H(\mathbb{B}^n)$, let

$$\delta(V, W) = \sup_{f \in V} \inf_{g \in W} \delta(f, g).$$

Also, let ρ be the Hausdorff metric on $H(\mathbb{B}^n)$ given by

$$\rho(V, W) = \max\{\delta(V, W), \delta(W, V)\},$$

for all nonempty compact subsets V and W of $H(\mathbb{B}^n)$.

In view of Theorem 2.7, the following result is related to [13, Proposition 4.20] and was obtained in [16] (cf. [23, Theorem I.45], for $n = 1$).

Proposition 4.2 *Let $A \in \widetilde{\mathcal{A}}$. Then $T \mapsto \widetilde{S}_A^T(\mathbb{B}^n)$ is a continuous mapping on $[0, \infty)$ to the metric space of nonempty compact subsets of $H(\mathbb{B}^n)$ with respect to ρ.*

The following remark is a consequence of Proposition 4.2 (see [16]).

Remark 4.3 In view of Proposition 4.2, we have that, for every $s \in (0, \infty]$ and $A \in \widetilde{\mathcal{A}}$, the following relations hold:

$$\widetilde{S}_A^0(\mathbb{B}^n) \subseteq \overline{\bigcup_{0 < t < s} \widetilde{S}_A^t(\mathbb{B}^n)} \quad \text{and} \quad \bigcap_{0 < t < s} \widetilde{S}_A^t(\mathbb{B}^n) \subseteq \widetilde{S}_A^0(\mathbb{B}^n).$$

The following remark points out a main difference between the case of time-dependent operators and that of time-independent operators (see [16]; cf. Remark 2.2 and Example 2.8).

Remark 4.4 For every $s > 0$, there exists

(i) $B \in \widetilde{\mathcal{A}}$ such that $\widetilde{S}_B^0(\mathbb{B}^2) \nsubseteq \bigcup_{t \geq s} \widetilde{S}_B^t(\mathbb{B}^2)$.
(ii) $B \in \widetilde{\mathcal{A}}$ such that $\bigcap_{t \geq s} \widetilde{S}_B^t(\mathbb{B}^2) \nsubseteq \widetilde{S}_B^0(\mathbb{B}^2)$.

In view of Remarks 4.3 and 4.4, it would be interesting to give an answer to the following question (see [16]):

Question 4.5 *Does there exist $A \in \widetilde{\mathcal{A}}$ such that $\widetilde{S}_A^0(\mathbb{B}^n) \nsubseteq \bigcup_{t > 0} \widetilde{S}_A^t(\mathbb{B}^n)$, for $n \geq 2$?*

Next, we consider the dependence of $\widetilde{S}_A^T(\mathbb{B}^n)$ on $A \in \widetilde{\mathcal{A}}$, when $T \geq 0$ is fixed (see [15]). Note that the following result may be seen as a *dominated convergence* type theorem for families of mappings with generalized parametric representation on $\mathbb{B}^n$ with respect to time-dependent operators $A \in \widetilde{\mathcal{A}}$.

Theorem 4.6 *Let $T \geq 0$ and $A \in \widetilde{\mathcal{A}}$ be such that $\operatorname{ess\,inf}_{t \geq T} m(A(t)) > 0$. Also, let $M > 0$, $\alpha \in L^1([T, \infty), \mathbb{R})$ and $(A_k)_{k \in \mathbb{N}}$ be a sequence in $\widetilde{\mathcal{A}}$ such that, for a.e. $t \geq T$ and for every $k \in \mathbb{N}$, we have: $\|A_k(t)\| \leq M$ and $\|V_k(T, t)^{-1}\| e^{-2 \int_T^t m(A_k(\tau))d\tau} \leq \alpha(t)$, where $V_k(T, \cdot)$ is the unique solution on $[T, \infty)$ of the initial value problem (1.1) related to A_k. If $A_k(t) \to A(t)$, as $k \to \infty$, for a.e. $t \geq T$, then $\rho(\widetilde{S}_{A_k}^T(\mathbb{B}^n), \widetilde{S}_A^T(\mathbb{B}^n)) \to 0$, as $k \to \infty$.*

In the case of time-independent operators $A \in L(\mathbb{C}^n)$ (cf. Remark 1.4), we have the following convergence results, due to Hamada, Iancu, Kohr [15]. These results are in connection with the perturbation theory of linear operators.

Theorem 4.7 *Let $\mathbf{A} \in L(\mathbb{C}^n)$ be such that $k_+(\mathbf{A}) < 2m(\mathbf{A})$, and let $(\mathbf{A}_l)_{l \in \mathbb{N}}$ be a sequence in $L(\mathbb{C}^n)$ such that $\mathbf{A}_l \to \mathbf{A}$, as $l \to \infty$. Then there is $l_0 \in \mathbb{N}$ such that $k_+(\mathbf{A}_l) < 2m(\mathbf{A}_l)$, $l \geq l_0$ (in particular, $S_{\mathbf{A}_l}^0(\mathbb{B}^n)$ and $\widehat{S}_{\mathbf{A}_l}(\mathbb{B}^n)$ are compact for $l \geq l_0$), $\rho(S_{\mathbf{A}_l}^0(\mathbb{B}^n), S_{\mathbf{A}}^0(\mathbb{B}^n)) \to 0$ and $\rho(\widehat{S}_{\mathbf{A}_l}(\mathbb{B}^n), \widehat{S}_{\mathbf{A}}(\mathbb{B}^n)) \to 0$, as $l \to \infty$.*

In view of Theorem 4.7, [1] and [28], the authors in [15] have proposed the following questions:

Question 4.8 *Let $n \geq 2$. Is it possible to generalize Theorem 4.7 to the case of non-resonant linear operators?*

Question 4.9 *Let $n \geq 2$. Under the assumption of Theorem 4.7, is it true that:*

$$\lim_{l \to \infty} \rho(\overline{\operatorname{ex} S_{\mathbf{A}_l}^0(\mathbb{B}^n)}, \overline{\operatorname{ex} S_{\mathbf{A}}^0(\mathbb{B}^n)}) = 0, \quad \lim_{l \to \infty} \rho(\overline{\operatorname{supp} S_{\mathbf{A}_l}^0(\mathbb{B}^n)}, \overline{\operatorname{supp} S_{\mathbf{A}}^0(\mathbb{B}^n)}) = 0?$$

In particular, is it true that:

$$\lim_{l \to \infty} \rho(\overline{\operatorname{ex} \widehat{S}_{\mathbf{A}_l}(\mathbb{B}^n)}, \overline{\operatorname{ex} \widehat{S}_{\mathbf{A}}(\mathbb{B}^n)}) = 0, \quad \lim_{l \to \infty} \rho(\overline{\operatorname{supp} \widehat{S}_{\mathbf{A}_l}(\mathbb{B}^n)}, \overline{\operatorname{supp} \widehat{S}_{\mathbf{A}_l}(\mathbb{B}^n)}) = 0?$$

The following convergence result related to the Carathéodory family $\mathcal{N}_\mathbf{A}$ was obtained in [15].

Theorem 4.10 *Let $\mathbf{A} \in L(\mathbb{C}^n)$ be such that $m(\mathbf{A}) > 0$, and let $(\mathbf{A}_k)_{k \in \mathbb{N}}$ be a sequence in $L(\mathbb{C}^n)$ such that $\mathbf{A}_k \to \mathbf{A}$, as $k \to \infty$. Then there is $k_0 \in \mathbb{N}$ such that $m(\mathbf{A}_k) > 0$, for all $k \geq k_0$, and $\rho(\mathcal{N}_{\mathbf{A}_k}, \mathcal{N}_\mathbf{A}) \to 0$, as $k \to \infty$.*

In view of Theorem 4.10, it is natural to ask the following question (cf. [15]):

Question 4.11 *Let $n \geq 2$. Under the assumption of Theorem 4.10, is it true that:*

$$\lim_{l \to \infty} \rho(\overline{\operatorname{ex} \mathcal{N}_{\mathbf{A}_l}}, \overline{\operatorname{ex} \mathcal{N}_\mathbf{A}}) = 0 \quad \text{and} \quad \lim_{l \to \infty} \rho(\overline{\operatorname{supp} \mathcal{N}_{\mathbf{A}_l}}, \overline{\operatorname{supp} \mathcal{N}_\mathbf{A}}) = 0?$$

Acknowledgements H. Hamada is partially supported by JSPS KAKENHI Grant Number JP16K05217.

References

1. Arosio, L.: Resonances in Loewner equations. Adv. Math. **227**, 1413–1435 (2011)
2. Bracci, F.: Shearing process and an example of a bounded support function in $S^0(\mathbb{B}^2)$. Comput. Methods Funct. Theory **15**, 151–157 (2015)
3. Bracci, F., Roth, O.: Support points and the Bieberbach conjecture in higher dimension. Preprint (2016); arXiv: 1603.01532v2
4. Bracci, F., Graham, I., Hamada, H., Kohr, G.: Variation of Loewner chains, extreme and support points in the class S^0 in higher dimensions. Constr. Approx. **43**, 231–251 (2016)
5. Dunford, N., Schwartz, J.: Linear Operators. Part I: General Theory. Interscience Publishers Inc., New York (1958)
6. Duren, P., Graham, I., Hamada, H., Kohr, G.: Solutions for the generalized Loewner differential equation in several complex variables. Math. Ann. **347**, 411–435 (2010)
7. Elin, M., Reich, S., Shoikhet, D.: Complex Dynamical Systems and the Geometry of Domains in Banach Spaces. Diss. Math. **427**, 1–62 (2004)
8. Graham, I., Kohr, G.: Geometric Function Theory in One and Higher Dimensions. Marcel Dekker, New York (2003)
9. Graham, I., Hamada, H., Kohr, G.: Parametric representation of univalent mappings in several complex variables. Can. J. Math. **54**, 324–351 (2002)
10. Graham, I., Hamada, H., Kohr, G., Kohr, M.: Asymptotically spirallike mappings in several complex variables. J. Anal. Math. **105**, 267–302 (2008)
11. Graham, I., Hamada, H., Kohr, G., Kohr, M.: Spirallike mappings and univalent subordination chains in $\mathbb{C}^n$. Ann. Scuola Norm. Sup. Pisa Cl. Sci. **7**, 717–740 (2008)
12. Graham, I., Hamada, H., Kohr, G., Kohr, M.: Extreme points, support points and the Loewner variation in several complex variables. Sci. China Math. **55**, 1353–1366 (2012)
13. Graham, I., Hamada, H., Kohr, G., Kohr, M.: Extremal properties associated with univalent subordination chains in $\mathbb{C}^n$. Math. Ann. **359**, 61–99 (2014)
14. Graham, I., Hamada, H., Kohr, G., Kohr, M.: Support points and extreme points for mappings with A-parametric representation in $\mathbb{C}^n$. J. Geom. Anal. **26**, 1560–1595 (2016)
15. Hamada, H., Iancu, M., Kohr, G.: Convergence results for families of univalent mappings on the unit ball in $\mathbb{C}^n$. Ann. Acad. Sci. Fenn. Math. **41**, 601–616 (2016)
16. Hamada, H., Iancu, M., Kohr, G.: Extremal problems for mappings with generalized parametric representation in $\mathbb{C}^n$. Complex Anal. Oper. Theory **10**, 1045–1080 (2016)
17. Kirwan, W.E.: Extremal properties of slit conformal mappings. In: Brannan, D., Clunie, J. (eds.) Aspects of Contemporary Complex Analysis, pp. 439–449. Academic Press, London (1980)
18. Pell, R.: Support point functions and the Loewner variation. Pac. J. Math. **86**, 561–564 (1980)
19. Pfaltzgraff, J.A.: Subordination chains and univalence of holomorphic mappings in $\mathbb{C}^n$. Math. Ann. **210**, 55–68 (1974)
20. Pommerenke, C.: Univalent Functions. Vandenhoeck and Ruprecht, Göttingen (1975)
21. Poreda, T.: On the univalent holomorphic maps of the unit polydisc in $\mathbb{C}^n$ which have the parametric representation, I-the geometrical properties. Ann. Univ. Mariae Curie Skl. Sect. A **41**, 105–113 (1987)
22. Poreda, T.: On generalized differential equations in Banach spaces. Diss. Math. **310**, 1–50 (1991)
23. Roth, O.: Control theory in $\mathcal{H}(\mathbb{D})$. Dissertation, Bayerischen University Wuerzburg (1998)
24. Roth, O.: Pontryagin's maximum principle for the Loewner equation in higher dimensions. Can. J. Math. **67**, 942–960 (2015)

25. Schleißinger, S.: On support points on the class $S^0(\mathbb{B}^n)$. Proc. Am. Math. Soc. **142**, 3881–3887 (2014)
26. Suffridge, T.J.: Starlikeness, convexity and other geometric properties of holomorphic maps in higher dimensions. In: Lecture Notes in Mathematics, vol. 599, pp. 146–159. Springer, Berlin (1977)
27. Voda, M.: Loewner theory in several complex variables and related problems. Ph.D. thesis, University of Toronto (2011)
28. Voda, M.: Solution of a Loewner chain equation in several complex variables. J. Math. Anal. Appl. **375**, 58–74 (2011)

Open Problems and New Directions for p-Modulus on Networks

Nathan Albin and Pietro Poggi-Corradini

Abstract The notion of p-modulus was created, and continues to play a vital role, in complex analysis and geometric function theory. Here we give an overview of the theory of p-modulus on networks that we have been developing in recent years. The hope is not only to develop a flexible tool on networks that can be useful for practical applications, but also that the rich unfolding theory on network will eventually inform the classical theory on metric measure spaces, Euclidean spaces, and the complex plane. We end by offering three open problems that are purely network theory problems. To keep the paper self-contained, we have not included possible applications, both to practical problems and to more theoretic function theoretic problems. We leave those directions for another time.

Keywords p-Modulus · Blocking duality

1991 Mathematics Subject Classification 90C35

1 Introduction

We begin by giving a brief review of the theory of p-modulus on networks. This is a subject rich with possibilities: the open directions are far too many to describe. Then, in the next section, we describe three possible open problems as a sample of the type of questions that can be pursued in this context.

N. Albin · P. Poggi-Corradini (✉)
Department of Mathematics, Kansas State University, Manhattan, KS, USA
e-mail: pietro@math.ksu.edu

© Springer International Publishing AG, part of Springer Nature 2017
F. Bracci (ed.), *Geometric Function Theory in Higher Dimension*,
Springer INdAM Series 26, https://doi.org/10.1007/978-3-319-73126-1_10

1.1   Modulus in the Continuum

The theory of conformal modulus was originally developed in complex analysis, see Ahlfors's comment on p. 81 of [1]. The more general theory of p-modulus grew out of the study of quasiconformal maps, which generalize the notion of conformal maps to higher dimensional real Euclidean spaces and, in fact, to abstract metric measure spaces. Intuitively, p-modulus provides a method for quantifying the richness of a family of curves, in the sense that a family with many short curves will have a larger modulus than a family with fewer and longer curves. The parameter p tends to favor the "many curves" aspect when p is close to 1 and the "short curves" aspect as p becomes large. This phenomenon was explored more precisely in [2] in the context of networks. he concept of discrete modulus on networks is not new, see for instance [8, 11, 14, 20]. However, recently the authors have started developing the theory of p-modulus as a graph-theoretic quantity [2, 5], with the goal of finding applications, for instance to the study of epidemics [13, 22].

As motivation for the discrete theory, let us recall the relevant definitions from the continuum theory. For now, it is convenient to restrict attention to the 2-modulus of curves in the plane, which, as it happens, is a conformal invariant and thus has been carefully studied in the literature.

Let Ω be a domain in $\mathbb{C}$, and let E, F be two continua in $\overline{\Omega}$. Define $\Gamma = \Gamma_\Omega(E, F)$ to be the family of all rectifiable curves connecting E to F in Ω. A *density* is a Borel measurable function $\rho : \Omega \to [0, \infty)$. We say that ρ is *admissible* for Γ and write $\rho \in \text{Adm}(\Gamma)$, if

$$\int_\gamma \rho \, ds \geq 1 \qquad \forall \gamma \in \Gamma, \tag{1.1}$$

where ds is arc-length. Now, we define the modulus of Γ as

$$\text{Mod}_2(\Gamma) := \inf_{\rho \in \text{Adm}(\Gamma)} \int_\Omega \rho^2 dA, \tag{1.2}$$

where dA is two-dimensional Lebesgue measure.

Example 1.1 (The Rectangle) Consider a rectangle

$$\Omega := \{z = x + iy \in \mathbb{C} : 0 < x < L, 0 < y < H\}$$

of height H and length L. Set $E := \{z \in \overline{\Omega} : \text{Re}\, z = 0\}$ and $F := \{z \in \overline{\Omega} : \text{Re}\, z = L\}$ to be the leftmost and rightmost vertical sides respectively. If $\Gamma = \Gamma_\Omega(E, F)$ then,

$$\text{Mod}_2(\Gamma) = \frac{H}{L}. \tag{1.3}$$

A famous and very useful result in this context is the notion of a *conjugate family* of a connecting family. For instance, in the case of the rectangle, the conjugate family $\Gamma^* = \Gamma_\Omega^*(E, F)$ for $\Gamma_\Omega(E, F)$ consists of all curves that "block" or intercept every curve $\gamma \in \Gamma_\Omega(E, F)$. It's clear in this case that Γ^* is also a connecting family, namely it includes every curve connecting the two horizontal sides of Ω. In particular, by (1.3), we must have $\text{Mod}_2(\Gamma^*) = L/H$. So we deduce that

$$\text{Mod}_2(\Gamma_\Omega(E, F)) \cdot \text{Mod}_2(\Gamma_\Omega^*(E, F)) = 1. \tag{1.4}$$

One reason this reciprocal relation is useful is that upper-bounds for modulus are fairly easy to obtain by choosing reasonable admissible densities and computing their energy. However, lower-bounds are typically harder to obtain. However, when an equation like (1.4) holds, then upper-bounds for the modulus of the conjugate family translate to lower-bounds for the given family.

In higher dimensions, say in $\mathbb{R}^3$, the conjugate family of a connecting family of curves consists of a family of surfaces, and therefore one must consider the concept of *surface modulus*, see for instance [18] and references therein. It is also possible to generalize the concept of modulus by replacing the exponent 2 in (1.2) with $p \geq 1$ and by replacing dA with a different measure.

In the paper [6] we establish a conjugate duality formula similar to (1.4) for p-modulus on networks, which we call *blocking duality*. This will be explained below, but first we will review some basic network theory and see how modulus on networks fits in that context.

1.2 Modulus on Networks

A general framework for modulus of objects on networks was developed in [4]. In what follows, $G = (V, E, \sigma)$ is taken to be a finite graph with vertex set V and edge set E. The graph may be directed or undirected and need not be simple. In general, we shall assume that each edge is assigned a corresponding weight $0 < \sigma(e) < \infty$.

We consider any finite family of "objects" Γ for which each $\gamma \in \Gamma$ can be assigned an associated function $\mathcal{N}(\gamma, \cdot) : E \to \mathbb{R}_{\geq 0}$ that measures the *usage of edge e by γ*. In the following it will be useful to define the quantity:

$$\mathcal{N}_{\min} := \min_{\gamma \in \Gamma} \min_{e : \mathcal{N}(\gamma, e) \neq 0} \mathcal{N}(\gamma, e). \tag{1.5}$$

We say Γ is *non-trivial* if $\mathcal{N}_{\min} > 0$.

Here are some examples of objects and their associated usage functions.

- A walk $\gamma = x_0\, e_1\, x_1 \cdots e_n\, x_n$ is associated to $\mathcal{N}(\gamma, e) =$ number times γ traverses e. In this case $\mathcal{N}(\gamma, \cdot) \in \mathbb{Z}_{\geq 0}^E$.
- A subset of edges $T \subset E$ is associated to its indicator function $\mathcal{N}(T, e) = \mathbb{1}_T(e) = 1$ if $e \in T$ and 0 otherwise. Here, $\mathcal{N}(\gamma, \cdot) \in \{0, 1\}^E$.

- A flow f is associated to its volume function $\mathcal{N}(f, e) = |f(e)|$. Therefore, $\mathcal{N}(\gamma, \cdot) \in \mathbb{R}^{E}_{\geq 0}$.

The matrix $\mathcal{N}$ is called the *usage matrix* for the family Γ. Each row of $\mathcal{N}$ corresponds to an object $\gamma \in \Gamma$ and records the usage of edge e by γ.

Note, that the families Γ under consideration may very well be infinite (e.g. families of walks), so $\mathcal{N}$ may have infinitely many rows. For this paper, we shall assume Γ is finite. This assumption is not quite as restrictive as it might seem. In [5] it was shown that any family Γ with an integer-valued $\mathcal{N}$ can be replaced, without changing the modulus, by a finite subfamily. For example, if Γ is the set of all walks between two distinct vertices, the modulus can be computed by considering only simple paths. This result implies a similar finiteness result for any family Γ whose usage matrix $\mathcal{N}$ is rational with positive entries bounded away from zero.

We define a *density* on G to be a nonnegative function on the edge set: $\rho : E \to [0, \infty)$. The value $\rho(e)$ can be thought of as the *cost of using edge e*. For an object $\gamma \in \Gamma$, we define

$$\ell_\rho(\gamma) := \sum_{e \in E} \mathcal{N}(\gamma, e)\rho(e) = (\mathcal{N}\rho)(\gamma),$$

which represents the *total usage cost* for γ with the given edge costs ρ. As in the continuum case, a density $\rho \in \mathbb{R}^{E}_{\geq 0}$ is called *admissible for* Γ, if

$$\ell_\rho(\gamma) \geq 1, \ \forall \gamma \in \Gamma, \text{ or equivalently, if } \ell_\rho(\Gamma) := \inf_{\gamma \in \Gamma} \ell_\rho(\gamma) \geq 1.$$

In matrix notation, ρ is admissible if $\mathcal{N}\rho \geq \mathbf{1}$, where $\mathbf{1}$ is the column vector of ones and the inequality is understood to hold elementwise. We define the set

$$\mathrm{Adm}(\Gamma) = \left\{\rho \in \mathbb{R}^{E}_{\geq 0} : \mathcal{N}\rho \geq \mathbf{1}\right\} \tag{1.6}$$

to be the set of admissible densities.

Now, for $1 \leq p \leq \infty$ we define the *p-energy* of a density ρ as

$$\mathcal{E}_{p,\sigma}(\rho) := \sum_{e \in E} \sigma(e)\rho(e)^p \quad \text{and} \quad \mathcal{E}_{\infty,\sigma}(\rho) := \max_{e \in E} \sigma(e)\rho(e)$$

Definition 1.2 Given a graph $G = (V, E, \sigma)$, a finite non-trivial family of objects Γ, and an exponent $1 \leq p \leq \infty$, then the *p-modulus* of Γ is

$$\mathrm{Mod}_{p,\sigma}(\Gamma) := \inf_{\rho \in \mathrm{Adm}(\Gamma)} \mathcal{E}_{p,\sigma}(\rho) \tag{1.7}$$

Remark 1.3 A extremal density ρ^* always exists for $1 \leq p \leq \infty$, satisfies $0 \leq \rho^* \leq \mathcal{N}^{-1}_{\min}$, and is unique for $1 < p < \infty$, see [2].

Modulus satisfies the following two basic properties:

- Γ-*monotonicity*:

$$\Gamma \subset \Gamma' \implies \text{Mod}_{p,\sigma}(\Gamma) \leq \text{Mod}_{p,\sigma}(\Gamma'), \tag{1.8}$$

for all $1 \leq p \leq \infty$, since $\text{Adm}(\Gamma') \subset \text{Adm}(\Gamma)$.
- *Subadditivity*:

$$\text{Mod}_{p,\sigma}\left(\bigcup_{j=1}^{\infty} \Gamma_j\right) \leq \sum_{j=1}^{\infty} \text{Mod}_{p,\sigma}(\Gamma_j). \tag{1.9}$$

1.3 Connection to Classical Quantities

The concept of p-modulus generalizes known classical ways of measuring the richness of a family of walks. Let $G = (V, E)$ and two vertices s and t in V be given. We define the *connecting family* $\Gamma(s, t)$ to be the family of all simple paths in G that start at s and end at t. A subset $C \subset E$ is called a *cut* for the family of paths Γ if for every $\gamma \in \Gamma$, there is $e \in C$ such that $\mathcal{N}(\gamma, e) = 1$. The size of a cut is measured by $|C| := \sum_{e \in C} \sigma(e)$. Also, in the undirected case, G can be thought of as an electrical network with edge conductances given by the weights σ, see [10]. We write $\mathcal{C}_{\text{eff}}(s, t)$ for the effective conductance between s and t. Here we summarize the various properties of continuity and monotonicity in p of modulus as well as the special cases for modulus of connecting families.

Theorem 1.4 ([2]) *Let $G = (V, E)$ be a graph with edge weights σ. Let Γ be a nontrivial family of objects on G with usage matrix $\mathcal{N}$ and let $\sigma(E) := \sum_{e \in E} \sigma(e)$. Then the function $p \mapsto \text{Mod}_{p,\sigma}(\Gamma)$ is continuous for $1 \leq p < \infty$, and the following two monotonicity properties hold for $1 \leq p \leq p' < \infty$.*

$$\mathcal{N}_{min}^{p} \text{Mod}_{p,\sigma}(\Gamma) \geq \mathcal{N}_{min}^{p'} \text{Mod}_{p',\sigma}(\Gamma), \tag{1.10}$$

$$\left(\sigma(E)^{-1} \text{Mod}_{p,\sigma}(\Gamma)\right)^{1/p} \leq \left(\sigma(E)^{-1} \text{Mod}_{p',\sigma}(\Gamma)\right)^{1/p'}. \tag{1.11}$$

Moreover, if $\Gamma = \Gamma(s, t)$ is a connecting family then

- ∞-*modulus recovers shortest-path*:

$$\text{Mod}_{\infty,1}(\Gamma) = \frac{1}{\ell(\Gamma)}.$$

- *1-modulus recovers mincut*:

$$\text{Mod}_{1,\sigma}(\Gamma) = \min\{|C| : C \text{ a cut of } \Gamma\}$$

- *2-modulus recovers effective conductance:*

$$\text{Mod}_{2,\sigma}(\Gamma) = \mathcal{C}_{\text{eff}}(s,t),$$

Remark 1.5 An early version of the case $p = 2$ is due to Duffin [11]. Also the $p = \infty$ case holds in general for any family Γ.

Example 1.6 (Basic Example) Let G be a graph consisting of k simple paths in parallel, each path taking ℓ hops to connect a given vertex s to a given vertex t. Assume also that G is unweighted, that is $\sigma \equiv 1$. Let Γ be the family consisting of the k simple paths from s to t. Then $\ell(\Gamma) = \ell$ and the size of the minimum cut is k. A straightforward computation shows that

$$\text{Mod}_p(\Gamma) = \frac{k}{\ell^{p-1}} \quad \text{for } 1 \le p < \infty, \qquad \text{Mod}_{\infty,1}(\Gamma) = \frac{1}{\ell}.$$

In particular, $\text{Mod}_p(\Gamma)$ is continuous in p, and $\lim_{p\to\infty} \text{Mod}_p(\Gamma)^{1/p} = \text{Mod}_{\infty,1}(\Gamma)$. Intuitively, when $p \approx 1$, $\text{Mod}_p(\Gamma)$ is more sensitive to the number of parallel paths, while for $p \gg 1$, $\text{Mod}_p(\Gamma)$ is more sensitive to short walks.

By Theorem 1.4, the concept of modulus encompasses classical quantities such as shortest path, minimal cut, and effective conductance. For this reason, modulus has many advantages. For instance, in order to give effective conductance a proper interpretation in terms of electrical networks one needs to consider the Laplacian operator which on undirected graphs is a symmetric matrix. On directed graphs however the Laplacian ceases to be symmetric, so the electrical network model breaks down. The definition of modulus, however, does not rely on the symmetry of the Laplacian and, therefore, can still be defined and computed in this case.

Moreover, modulus can measure the richness of many types of families of objects, not just connecting walk families. Here are some examples of families of objects that can be measured using modulus and that we have been actively investigating:

- **Spanning Tree Modulus:** All spanning trees of G, [3].
- **Loop Modulus:** All simple cycles in G, [21]
- **Via Modulus:** All walks that start at s, end at t, and visit u along the way, [22].

There are many more families that we are either in the process of investigating. For instance, families of perfect or full matchings, families of long paths (e.g., all simple paths with at least L hops), and the family of all the stars in a graph (a star is the set of edges that are incident at a vertex).

1.4 Blocking Duality and p-Modulus

It turns out that many families of objects have a naturally associated "dual" family of objects, which goes under the name of the blocker family.

First, we recall some general definition. Let $\mathcal{K}$ be the set of all closed convex sets $K \subset \mathbb{R}^E_{\geq 0}$ that are *recessive*, in the sense that $K + \mathbb{R}^E_{\geq 0} = K$. To avoid trivial cases, we shall assume that $\varnothing \subsetneq K \subsetneq \mathbb{R}^E_{\geq 0}$, for $K \in \mathcal{K}$. For each $K \in \mathcal{K}$ there is an associated *blocking polyhedron*, or *blocker*, $\mathrm{BL}(K) := \{\eta \in \mathbb{R}^E_{\geq 0} : \eta^T \rho \geq 1, \ \forall \rho \in K\}$. Given $K \in \mathcal{K}$ and a point $x \in K$ we say that x is an *extreme point* of K if $x = t x_1 + (1-t)x_2$ for some $x_1, x_2 \in K$ and some $t \in (0, 1)$, implies that $x_1 = x_2 = x$. Moreover, we let $\mathrm{ext}(K)$ be the set of all extreme points of K. The *dominant* of a set $P \subset \mathbb{R}^E_{\geq 0}$ is the recessive closed convex set $\mathrm{Dom}(P) = \mathrm{co}(P) + \mathbb{R}^E_{\geq 0}$.

If Γ is a finite non-trivial family of objects on a graph G, the admissible set $\mathrm{Adm}(\Gamma)$ is determined by finitely many inequalities, in particular $\mathrm{Adm}(\Gamma)$ has finitely many faces. However, $\mathrm{Adm}(\Gamma)$ is also determined by its finitely many extreme points, or "vertices". In fact, it is well-known that $\mathrm{Adm}(\Gamma)$ is the dominant of its extreme points $\mathrm{ext}(\mathrm{Adm}(\Gamma))$, see [19, Theorem 18.5].

Definition 1.7 Suppose $G = (V, E)$ is a finite graph and Γ is a finite non-trivial family of objects on G. We say that the family

$$\hat{\Gamma} := \mathrm{ext}(\mathrm{Adm}(\Gamma)) = \{\hat{\gamma}_1, \ldots, \hat{\gamma}_s\} \subset \mathbb{R}^E_{\geq 0},$$

consisting of the extreme points of $\mathrm{Adm}(\Gamma)$, is the *Fulkerson blocker* of Γ. Also, we define the matrix $\hat{\mathcal{N}} \in \mathbb{R}^{\hat{\Gamma} \times E}_{\geq 0}$ to be the matrix whose rows are the vectors $\hat{\gamma}^T$, for $\hat{\gamma} \in \hat{\Gamma}$.

Theorem 1.8 (Fulkerson [12]) *Let $G = (V, E)$ be a graph and let Γ be a non-trivial finite family of objects on G. We identify Γ with its edge-usage functions, hence we think of Γ as a finite subset of $\mathbb{R}^E_{\geq 0}$. Let $\hat{\Gamma}$ be the Fulkerson blocker of Γ. Then*

(1) $\mathrm{Adm}(\Gamma) = \mathrm{Dom}(\hat{\Gamma}) = \mathrm{BL}(\mathrm{Adm}(\hat{\Gamma}))$;
(2) $\mathrm{Adm}(\hat{\Gamma}) = \mathrm{Dom}(\Gamma) = \mathrm{BL}(\mathrm{Adm}(\Gamma))$;
(3) $\hat{\hat{\Gamma}} \subset \Gamma$.

In particular,

$$\mathrm{BL}(\mathrm{BL}(\mathrm{Adm}(\Gamma))) = \mathrm{Adm}(\Gamma) \quad and \quad \mathrm{BL}(\mathrm{BL}(\mathrm{Dom}(\Gamma))) = \mathrm{Dom}(\Gamma).$$

as well as

$$\mathrm{Adm}(\Gamma) = \mathrm{BL}(\mathrm{Dom}(\Gamma)) \quad and \quad \mathrm{BL}(\mathrm{Adm}(\Gamma)) = \mathrm{Dom}(\Gamma).$$

For a proof of Theorem 1.8 see [6].

1.5   Blocking Duality for p-Modulus

The following is a generalization on networks of the concept of conjugate families in complex analysis.

Theorem 1.9 ([6]) *Let $G = (V, E)$ be a graph and let Γ be a nontrivial finite family of objects on G with Fulkerson blocker $\hat{\Gamma}$. Let the exponent $1 < p < \infty$ be given, with $q = (p-1)/p$ its conjugate exponent. For any set of weights $\sigma \in \mathbb{R}_{>0}^E$ define the dual set of weights $\hat{\sigma}$ as $\hat{\sigma}(e) := \sigma(e)^{-\frac{q}{p}}$, for all $e \in E$.*
   Then

$$\mathrm{Mod}_{p,\sigma}(\Gamma)^{\frac{1}{p}} \, \mathrm{Mod}_{q,\hat{\sigma}}(\hat{\Gamma})^{\frac{1}{q}} = 1. \tag{1.12}$$

Moreover, the optimal $\rho^* \in \mathrm{Adm}(\Gamma)$ and $\eta^* \in \mathrm{Adm}(\hat{\Gamma})$ are unique and are related as follows:

$$\eta^*(e) = \frac{\sigma(e)\rho^*(e)^{p-1}}{\mathrm{Mod}_{p,\sigma}(\Gamma)} \qquad \forall e \in E. \tag{1.13}$$

Also

$$\mathrm{Mod}_{1,\sigma}(\Gamma) \, \mathrm{Mod}_{\infty,\sigma^{-1}}(\hat{\Gamma}) = 1. \tag{1.14}$$

Remark 1.10 The case for $p = 2$ is essentially contained in [16, Lemma 2], although not stated in terms of modulus and with a different proof. However, it is worth stating it separately. Namely,

$$\mathrm{Mod}_{2,\sigma}(\Gamma) \, \mathrm{Mod}_{2,\sigma^{-1}}(\hat{\Gamma}) = 1 \quad \text{and} \quad \sigma(e)\rho^*(e) = \mathrm{Mod}_{2,\sigma}(\Gamma)\eta^*(e) \qquad \forall e \in E, \tag{1.15}$$

1.6   The Probabilistic Interpretation of Modulus

For simplicity, assume that G is unweighted and let $p = 2$. Also assume Γ is finite of subsets of E. Then blocker duality gives rise to a probabilistic interpretation of modulus. The idea is to use the fact that $\mathrm{Mod}_2(\Gamma) = \mathrm{Mod}_2(\hat{\Gamma})^{-1}$ and that $\hat{\hat{\Gamma}} \subset \Gamma$. Then the $\mathrm{Mod}_2(\hat{\Gamma})$ problem can be phrased as follows:

$$\text{minimize} \quad \mathbb{E}_2(\eta)$$

$$\text{subject to} \quad \eta(e) = \mathbb{P}_\mu\left(e \in \underline{\gamma}\right) \, \forall e \in E \tag{1.16}$$

$$\mu \in \mathcal{P}(\Gamma),$$

where $\mathcal{P}$ is the set of probability measures on Γ.

Note that while an optimal η^* always exists and is unique, the corresponding *optimal measures* μ^* are not necessarily unique.

2 Open Problems and New Directions

2.1 Fulkerson Blocker Pairs

We propose a systematic study of Fulkerson blocker pairs. This is a topic that is well-studied in the literature in combinatorics, however, the result are not always phrased in our terminology of families of objects and admissible densities. Also, the proofs in this context sometime feel ad hoc, and it would be nice to have a more unified approach.

The first and classical blocker pair consists of $\Gamma(s, t)$ the family of connecting simple s/t-paths whose Fulkerson blockers is the family $\hat{\Gamma}(s, t)$ of minimal s/t-cuts. The proof of blocker duality in this case is based on the Maxflow-Mincut Theorem.

Another known pair is the family Γ_{spt} of all spanning trees, whose Fulkerson blocker is the family of feasible partitions appropriately normalized. A *feasible partition P* is a partition of the vertex set V into two or more subsets, $\{V_1, \ldots, V_{k_P}\}$, such that each of the induced subgraphs $G(V_i)$ is connected. The corresponding edge set, E_P, is defined to be the set of edges in G that connect vertices belonging to different V_i's. Results of [9] imply that if Γ is the family of spanning trees on G. Then the Fulkerson blocker of Γ is the set of all vectors $(k_P - 1)^{-1} \mathbb{1}_{E_P}$, ranging over all feasible partitions P.

In [16] Lovasz mentions many examples of blocker pairs, although with a different definition of blocker. For instance, when G is a bipartite graph the family Γ of all perfect matchings is dual to the family of all odd cuts. A *perfect matching* is a subgraph H of G such that every vertex in V has H-degree equal to 1. An *odd cut* is the edge boundary ∂S of a subset $S \subset V$ of odd cardinality.

2.2 Graph Metrics

Consider a simple graph $G = (V, E, \sigma)$. There are many known graph metrics that measure the distance between two nodes $s, t \in V$. We mention two:

- **Shortest-path:** This consists in minimizing the length $\ell_\sigma(\gamma)$ of any walk γ from s to t.
- **Effective resistance:** This involves thinking of $\sigma(e)$ as an electrical conductance and $r(e) = \sigma(e)^{-1}$ as electrical resistance. Then $\mathcal{R}_{\text{eff}}(s, t)$ is the potential drop needed at s and t in order to pass a unit current flow through the network from s to t.

As shown in Theorem 1.4, shortest path is related to ∞-modulus and effective resistance is related to 2-modulus. We make the following definition for $1 \leq p < \infty$:

$$d_p := \left(\mathrm{Mod}_p(\Gamma(s, t))\right)^{-1/p},$$

where $\Gamma(s, t)$ is the family of all walks connecting s to t.

In a recent preprint we have shown the following.

Proposition 2.1 *For $1 \leq p < \infty$, d_p is a metric on V. Moreover, d_1 is an ultrametric.*

Recall that d is an *ultrametric* if $d(a, b) \leq \max\{d(a, c), d(c, b)\}$ for every a, b, c. Also as $p \to \infty$, d_p recovers the shortest path metric.

When $p = 2$, $d_2 = \sqrt{\mathcal{R}_{\mathrm{eff}}}$ which is a metric that arises naturally in the theory of the Laplacian. However, if d is a metric, then d^t is also a metric for all $0 < t < 1$. So it is natural to ask to how high an exponent can a metric d be raised.

Conjecture 2.2 From early numerical experiments we conjecture that d_p^q is always a metric when $1 < p < \infty$ and $q := p/(p - 1)$ is the Hölder conjugate exponent of p.

Note that, when $p = 2$, $d_p^q = \mathcal{R}_{\mathrm{eff}}$, so we recover effective resistance. Also, when $p = 1$, d_1 is an ultrametric and so are all of its powers.

Considering the path graph P_3 on three nodes a, b, c with edges $\{a, b\}$ and $\{b, c\}$, we can see that d_p^q is indeed a metric in that case and that $d_p^{q+\epsilon}$ is not a metric for all $\epsilon > 0$. So this graphs show that if the conjecture is true, then it cannot be improved unless one restricts to special families of graphs.

Added in Proof After this paper was submitted we were able to prove Conjecture 2.2 in the affirmative. The proof is based on Fulkerson duality and hence was added to [6]. This also gives a new modulus-based proof of the fact that effective resistance is a metric on graphs, which we will now quickly sketch. The fact that effective resistance $d_{\mathrm{ER}}(a, b) := \mathcal{R}_{\mathrm{eff}}(a, b)$ is a metric has several known proofs. See [15, Exercise 9.8], for a proof using current flows, and see [15, Corollary 10.8], for one using commute times. However, none of these other known proofs appear to generalize to give Conjecture 2.2 for $p \neq 2$. Our modulus proof, on the other hand, generalizes easily once the tool of Fulkerson duality (Theorem 1.9) is established.

Let $G = (V, E)$ be a simple connected graph. We want to prove that effective resistance is a metric. For simplicity we will only check that the triangle inequality holds when a, b, c are three distinct vertices in V. First, by Duffin's original result (see Theorem 1.4),

$$\mathcal{R}_{\mathrm{eff}}(s, t) = \mathrm{Mod}_2(\Gamma(s, t))^{-1},$$

for any pair of nodes $s, t \in V$. Next, by Fulkerson duality (Theorem 1.9):

$$\mathrm{Mod}_2(\Gamma(s, t))^{-1} = \mathrm{Mod}_2(\hat{\Gamma}(s, t)),$$

where $\hat{\Gamma}(s, t)$ is the family of minimal s/t-cuts. So it will be enough to show that

$$\text{Mod}_2(\hat{\Gamma}(a, b)) \le \text{Mod}_2(\hat{\Gamma}(a, c)) + \text{Mod}_2(\hat{\Gamma}(c, b)). \tag{2.1}$$

Notice that $\text{Mod}_2(\hat{\Gamma}(s, t))$ does not change if we enlarge $\hat{\Gamma}(s, t)$ to include cuts that may not be minimal (because every cut contains a minimal one, so every density that is admissible on the minimal cuts, will also be admissible on the enlarged family). Now fix an a/b-cut $S \subset V$, so that $a \in S$ and $b \notin S$. Then either $c \in S$ or $c \notin S$. Namely

$$\hat{\Gamma}(a, b) \subset \hat{\Gamma}(a, c) \cup \hat{\Gamma}(c, b).$$

Therefore (2.1) follows from the basic monotonicity property (1.8) and subadditivity property (1.9) of modulus.

2.3 Optimal Spanning Trees

Given a connected, unweighted, multigraph $G = (V, E)$ (with no self-loops), let Γ be the family of all spanning trees. In general, Γ is a very large family. For instance, for the complete graph $G = K_N$ on N nodes, Cayley's Theorem says that $|\Gamma| = N^{N-2}$. In general, Kirchhoff's Matrix Theorem gives a beautiful formula to compute $|\Gamma|$ in terms of the non-zero eigenvalues of the Laplacian of G. We are interested in computing the modulus of Γ and studying the families of *optimal trees*, namely the trees that are in the support of an optimal measure μ^*, see Sect. 1.6.

For simplicity we fix $p = 2$ in this discussion (although it turns out that what we are about to say does not depend on p). To begin with, recall that a vertex whose removal disconnects G is called an *articulation point* and G is said to be *biconnected* if it has no articulation points. More generally, G is *k-connected* if removing $k - 1$ vertices does not disconnect the graph. Moreover, every connected multigraph has a unique decomposition into maximal biconnected components which meet at articulation points.

The first thing we show in [7] is that spanning tree modulus obeys a sort of "serial" rule. Suppose G has maximal biconnected components $G_1, G_2, \ldots, G_k$. Let η_i^* be the optimal density for the dual problem on G_i, and let μ_i^* be a corresponding optimal pmf. Define $\mu \in \mathcal{P}(\Gamma_G)$ so that, for every tree $\gamma = \gamma_1 \cup \gamma_2 \cup \cdots \cup \gamma_k$, with $\gamma_i \in \Gamma_{G_i}$ for $i = 1, \ldots, k$,

$$\mu(\gamma) := \prod_{i=1}^k \mu_i^*(\gamma_i), \quad \text{and let} \quad \eta(e) := \mathbb{P}_\mu\left(e \in \underline{\gamma}\right).$$

Then η is optimal for the dual problem (1.16) on G, $\eta(e) = \eta_i^*(e)$ whenever $e \in E_{G_i}$, and the following serial rule holds:

$$\mathrm{Mod}_2(\Gamma_G) = \left(\sum_{i=1}^k \mathrm{Mod}_2(\Gamma_{G_i})^{-1} \right)^{-1}.$$

Therefore, without loss of generality, we can restrict our attention to biconnected graphs. More importantly, we define the notion of *homogeneous graphs*. These are the graphs for which the optimal density ρ^* and hence also the optimal edge-probabilities η^* (see (1.16)) are constant. Homogeneous graphs have several interesting properties, for instance, more that $|E|/(|V|-1)$ edges must be removed in order to disconnect them. Our main result is that every graph contains a non-trivial homogeneous core, i.e., a vertex induced subgraph H of G that is homogeneous with the property that every optimal tree for G restricts to an optimal tree for H. This triggers a *deflation process* where we repeatedly find the homogeneous core and shrink it to a single vertex hence yielding a decomposition of the original graph G into homogeneous components.

The upshot, due to the restriction property of optimal trees, is that we can from now on assume without loss of generality that G is a homogeneous graph. The question now is how to understand optimal trees on homogeneous graphs. First recall that uniform spanning trees are related to effective resistance by a famous result of Kirchhoff, see [17]. More specifically, if μ_0 is the uniform measure on Γ, then the resulting edge-probabilities satisfy

$$\mathbb{P}_{\mu_0}\left(e \in \underline{\gamma} \right) = \mathcal{R}_{\mathrm{eff}}(e).$$

Also, there are nice algorithms to produce uniform spanning trees, such as Wilson's algorithm and Aldous' algorithm.

Conjecture 2.3 Every homogeneous graph can be further decomposed into graphs that are uniform for an appropriate choice of edge-weights.

If this conjecture were to hold, then the study of optimal trees for modulus would reduce to the study of uniform spanning trees on pieces of the original graph.

Acknowledgements The authors are supported by NSF n. 1201427 and n. 1515810.

References

1. Ahlfors, L.V.: Conformal Invariants: Topics in Geometric Function Theory. McGraw-Hill, New York (1973)
2. Albin, N., Brunner, M., Perez, R., Poggi-Corradini, P., Wiens, N.: Modulus on graphs as a generalization of standard graph theoretic quantities. Conform. Geom. Dyn. **19**, 298–317 (2015). https://doi.org/10.1090/ecgd/287

3. Albin, N., Clemens, J., Hoare, D., Poggi-Corradini, P., Sit, B., Tymochko, S.: Spanning tree modulus: homogeneous graphs and deflation (2016). Preprint
4. Albin, N., Poggi-Corradini, P.: Minimal subfamilies and the probabilistic interpretation for modulus on graphs. J. Anal. 1–26 (2016). https://doi.org/10.1007/s41478-016-0002-9
5. Goering, M., Albin, N., Sahneh, F., Scoglio, C., Poggi-Corradini, P.: Numerical investigation of metrics for epidemic processes on graphs. In: 49th Asilomar Conference on Signals, Systems and Computers, pp. 1317–1322 (2015)
6. Albin, N., Clemens, J., Fernando, N., Poggi-Corradini, P.: Blocking duality for p-modulus on networks and applications. Preprint(2017). arXiv:1612.00435
7. Albin, N., Clemens, J., Hoare, D., Poggi-Corradini, P., Sit, B., Tymochko, S.: Minimizing the expected overlap of random spanning trees on graphs (2018). Preprint
8. Cannon, J.W., Floyd, W.J., Parry, W.R.: Squaring rectangles: the finite Riemann mapping theorem. In: The Mathematical Legacy of Wilhelm Magnus: Groups, Geometry and Special Functions (Brooklyn, NY, 1992). Contemporary Mathematics, vol. 169, pp. 133–212. American Mathematical Society, Providence, RI (1994)
9. Chopra, S.: On the spanning tree polyhedron. Oper. Res. Lett. **8**(1), 25–29 (1989)
10. Doyle, P.G., Snell, J.L.: Random walks and electric networks, Carus Mathematical Monographs, vol. 22. Mathematical Association of America, Washington, DC (1984)
11. Duffin, R.J.: The extremal length of a network. J. Math. Anal. Appl. **5**(2), 200–215 (1962)
12. Fulkerson, D.R.: Blocking polyhedra. Technical report, DTIC Document (1968)
13. Goering, M., Albin, N., Sahneh, F., Scoglio, C., Poggi-Corradini, P.: Numerical investigation of metrics for epidemic processes on graphs. In: 2015 49th Asilomar Conference on Signals, Systems and Computers, pp. 1317–1322 (2015). https://doi.org/10.1109/ACSSC.2015.7421356
14. Haïssinsky, P.: Empilements de cercles et modules combinatoires. Annales de l'Institut Fourier **59**(6), 2175–2222 (2009).Version revisée et corrigée
15. Levin, D.A., Peres, Y., Wilmer, E.L.: Markov Chains and Mixing Times. American Mathematical Society, Providence, RI (2009). With a chapter by James G. Propp and David B. Wilson
16. Lovász, L.: Energy of convex sets, shortest paths, and resistance. J. Comb. Theory Ser. A **94**(2), 363–382 (2001)
17. Lyons, R., Peres, Y.: Probability on Trees and Networks. Cambridge University Press, New York (2016). Available at http://pages.iu.edu/~rdlyons/
18. Rajala, K.: The local homeomorphism property of spatial quasiregular mappings with distortion close to one. Geom. Funct. Anal. **15**(5), 1100–1127 (2005)
19. Rockafellar, R.T.: Convex Analysis. Princeton University Press, Princeton (1970)
20. Schramm, O.: Square tilings with prescribed combinatorics. Isr. J. Math. **84**(1-2), 97–118 (1993)
21. Shakeri, H., Poggi-Corradini, P., Albin, N., Scoglio, C.: Network clustering and community detection using modulus of families of loops (2016). https://arxiv.org/pdf/1609.00461.pdf
22. Shakeri, H., Poggi-Corradini, P., Scoglio, C., Albin, N.: Generalized network measures based on modulus of families of walks. J. Comput. Appl. Math. (2016). http://dx.doi.org/10.1016/j.cam.2016.01.027

Metric Properties of Domains in $\mathbb{C}^n$

Hervé Gaussier

Abstract This article is the written form of a presentation given during the Workshop "Geometric Function Theory in higher dimension". We present some developments in the classical field of classification of domains in $\mathbb{C}^n$ and some related questions.

Keywords Complex manifolds · Automorphism groups · Kobayashi hyperbolicity · Horospheres · D'Angelo finite type

2010 Mathematics Subject Classification 32M99, 32Q45

1 Introduction

This article is an attempt to present several questions in the classical area of Several Complex Variables. Although most of them are relevant in the context of complex manifolds, we will focus essentially on domains in the complex Euclidean space $\mathbb{C}^n$. The problem of classifying complex manifolds is the starting point of the different questions, this is also their common thread. The article collects different questions raised in the purpose to generalize the Riemann Mapping Theorem for domains in $\mathbb{C}^n$; only very special cases are understood. They are at the intersection between the function theory and the complex geometry of complex manifolds, the metric geometry of invariant distances and the Cauchy-Riemann geometry of submanifolds of complex manifolds. By combining concepts and methods from that different fields, one might hope to approach long-standing classical problems in Several Complex Variables.

H. Gaussier (✉)
Université Grenoble Alpes, CNRS, Grenoble, France
e-mail: herve.gaussier@univ-grenoble-alpes.fr

© Springer International Publishing AG, part of Springer Nature 2017
F. Bracci (ed.), *Geometric Function Theory in Higher Dimension*,
Springer INdAM Series 26, https://doi.org/10.1007/978-3-319-73126-1_11

As expected, there is no possible complete classification of complex manifolds in complex dimension greater than or equal to 2. By imposing geometric and dynamical conditions on the manifold (after all, the unit disk in $\mathbb{C}$ is a homogeneous complex manifold), different classes of manifolds were exhibited, with models in the vein of the unit disk. However, such result are rare and many questions remain open in that direction. These questions concern not only the classification problem but also metric properties of complex manifolds. These are not necessarily major issues, but they are important ingredients that constituted approaches to the classification problem and that provide a better understanding of the complex geometry of open complex manifolds with boundaries. We will neither consider Kähler metrics in open manifolds with boundary, nor present in details the CR classification of CR manifolds, although these might be major subjects. We will focus mainly on Kobayashi hyperbolic manifolds. For a complete presentation of hyperbolic manifolds, with applications, see the book [22] by Kobayashi. We quoted only very few results in the article, forgetting many strong results obtained in the different questions mentioned here. We apologize in advance to the different authors for not having mentioned all the works related to the subject.

2   Classification of Domains in $\mathbb{C}^n$

The question is the following: given two complex manifolds, are they holomorphically equivalent, i.e. does there exist a biholomorphism between them? The Uniformization Theorem provides a complete classification in complex dimension one: any non-empty simply connected open subset of a Riemann surface is biholomorphic to one of the following: the Riemann sphere, the complex line $\mathbb{C}$ or the unit disk $\Delta := \{z \in \mathbb{C} : |z| < 1\}$. This theorem gives a purely topological classification: if two such open sets are topologically equivalent, then they are holomorphically equivalent. In particular, every simply connected domain in $\mathbb{C}$, different from $\mathbb{C}$, is biholomorphic to the unit disk. That result, known as the Riemann Mapping Theorem, was first stated by B. Riemann in 1851, although seemingly proved much later. It implies that every such domain is homogeneous, meaning that its automorphism group is transitive. Here *Automorphism group of a domain* means the group of self biholomorphisms of the domain.

The field of several complex variables developed much later, probably due to the lack of striking specific results, not covered by the one dimensional case. The real development of the field started essentially in 1895, with the Weierstrass Preparation Theorem. Concerning the classification of complex manifolds, the first striking result that gained high attention is a negative result due to Poincaré [25], in 1907:

Theorem 2.1 *There is no biholomorphism between the unit ball* $\mathbb{B}^2 = \{z \in \mathbb{C}^2 : ||z|| < 1\}$ *and the polydisk* $\Delta \times \Delta$.

This result shows a conceptual difference between Complex Geometry in complex dimension one and Complex Geometry in higher complex dimension: the classification in higher dimension cannot be reduced to a topological classification and different (complex) geometric structures can be attached to boundaries of open manifolds. The proof presented by Poincaré is essentially algebraic: if these two homogeneous domains were equivalent, there would exist a biholomorphism $f : \mathbb{B}^2 \to \Delta \times \Delta$, satisfying $f(0) = 0$. Hence the automorphism groups $Aut(\mathbb{B}^2)$ and $Aut(\Delta \times \Delta)$ would be isomorphic, and the same property would hold for their connected components of identity (we recall that $Aut(\mathbb{B}^2)$ and $Aut(\Delta \times \Delta)$ are topological groups for the compact open topology). However, the connected component of Identity of $Aut(\Delta \times \Delta)$ reduces to $\{(z, w) \mapsto (e^{i\theta}z, e^{i\eta}w)\}$ and hence is Abelian. However, the connected component of Identity of $Aut(\mathbb{B}^2)$ contains all unitary transforms in $\mathbb{C}^2$ and, hence, cannot be abelian. This gives a contradiction.

The obstruction for the holomorphic equivalence between $\mathbb{B}^2$ and $\Delta \times \Delta$ being not topological, obstructions (also called invariant) may be of different types. The structure of the automorphism group of a complex manifold being constrained by the geometry of the manifold, one may seek for geometric invariants. In Theorem 2.1, the obstruction comes from the complex structure of the boundaries: the unit sphere in $\mathbb{C}^2$ does not contain non trivial complex submanifolds, whereas the boundary of the polydisk is foliated by complex disks. Note that at this level, the obstruction is heuristic. One would expect, if there were a biholomorphism, that it would extend up to the boundaries as a homeomorphism and the contradiction would come from the different CR structures of the boundaries. However that approach relies on a fundamental property: the extension of biholomorphisms up to the boundary; this is a difficult question in general. We should point out that there is no one-to-one correspondence between the equivalence of CR structures of the boundaries of two complex manifolds and the holomorphic equivalence of the manifolds: there exist, for instance, domains not containing non trivial complex submanifolds in their boundaries that are biholomorphic to domains whose boundary contains non trivial complex submanifolds.

After Poincaré's Theorem, it is natural to study the classification of complex manifolds in complex dimension strictly larger than one. One may observe the following facts:

1. Let M and M' be two complex manifolds and let $f : M \to M'$ be a biholomorphism. Assume that their automorphism groups are topological groups (this is for instance the case if M and M' are bounded domains in $\mathbb{C}^n$). Then $Aut(M)$ is compact if and only if $Aut(M')$ is compact.
2. There are "more" smooth diffeomorphisms in $\mathbb{C}^n$ than biholomorphisms. Hence two generic domains are not holomorphically equivalent; generically, two small deformations of the unit ball have automorphism groups reduced to Identity but are not holomorphically equivalent.

In particular, there is no hope to obtain a general classification of complex manifolds and, in order to get some partial classification, one needs geometric or dynamical assumptions, such as the non compactness of the automorphism group. In case of a bounded domain D in $\mathbb{C}^n$, the non compactness of $Aut(D)$, for the compact open topology, is equivalent to the following:

$$\exists z^0 \in D, \ \exists p \in \partial D, \ \exists (\varphi_\nu)_\nu \in (Aut(D))^{\mathbb{N}} / \ \lim_{\nu \to \infty} \varphi_\nu(z^0) = p. \qquad (2.1)$$

We then say that p is an *accumulation point* for $Aut(D)$. Condition (2.1) remains a relevant assumption for not necessarily bounded domains with non compact automorphism group.

One might expect that, as soon as Condition (2.1) is satisfied, the global "inside" geometry of D might be determined by the local geometry of ∂D near p, under reasonable geometric assumptions on ∂D near p. This is the content of the Wong-Rosay Theorem:

Theorem 2.2 ([24, 26, 27]) *Let D be a domain in $\mathbb{C}^n$, $n \geq 2$. Let $p \in \partial D$ be an accumulation point for $Aut(D)$. If ∂D is of class C^2 and strictly pseudoconvex near p, then D is biholomorphic to the unit ball $\mathbb{B}^n$ in $\mathbb{C}^n$.*

The formulation of the Wong-Rosay Theorem presented here is due to S. Pinchuk: the domain may be unbounded and the geometric assumption on ∂D is local at p. We recall that ∂D is strictly pseudoconvex near a point $p \in \partial D$ if and only if there exists a neighborhood U of p in $\mathbb{C}^n$ and a biholomorphism $F : U \to F(U) \subset \mathbb{C}^n$ such that $\partial D \cap U$ is of class C^2 and $F(U \cap \overline{D})$ is strictly convex.

The proof of Theorem 2.2 is as follows, see Fig. 1. It is immediate, by the definition of strict pseudoconvexity, that there exists a local holomorphic peak function at p, namely a continuous function $h : U \cap \overline{D} \to \mathbb{C}$, holomorphic on $U \cap D$, such that $h(p) = 1$ and $|h(z)| < 1$ for every $z \in U \cap \overline{D} \backslash \{p\}$.

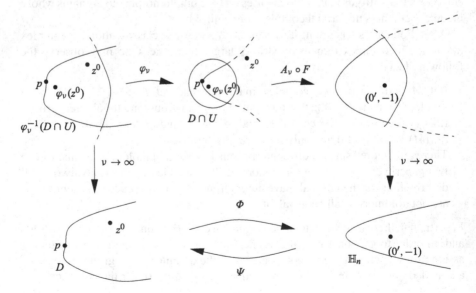

Fig. 1 presents the Scaling Process for strongly pseudoconvex domains

Claim 1 Let $z^0 \in D$, $p \in \partial D$, $(\varphi_\nu)_\nu \in (Aut(D))^{\mathbb{N}}$ be such that $\lim_{\nu \to \infty} \varphi_\nu(z^0) = p$. Then, for every $K \subset\subset D$, we have: $\lim_{\nu \to \infty} \varphi_\nu(K) = p$.

The proof of Claim 1 relies only on the existence of a holomorphic peak function at p (see for instance [5, 16] where the attraction property of holomorphic disks is considered).

Claim 2 For every $\nu \gg 1$, there exists an affine automorphism A_ν of $\mathbb{C}^n$ such that $A_\nu \circ F(\varphi_\nu(z^0)) = 0$ and, for every $0 \in K \subset\subset \mathbb{C}^n$, $A_\nu \circ F(U \cap \overline{D}) \cap K$ converges, for the Hausdorff convergence of sets, to the set $K \cap \mathbb{H}_n$ where $\mathbb{H}_n := \{z = (z', z_n) \in \mathbb{C}^n : Re(z_n) + ||z'||^2 < 0\}$ denotes the Heisengroup group, unbounded representation of $\mathbb{B}^n$.

Claim 2 is known as the method of dilations and was introduced by S. Pinchuk in Several Complex Variables. Since $F(U \cap \overline{D})$ is strictly convex and smooth, one can view this as a small deformation of $\mathbb{H}_n$ after a suitable change of coordinates. Then A_ν is a composition $A_\nu := \Lambda_\nu \circ U_\nu \circ T_\nu$, where T_ν is a translation sending the closest point to $F(\varphi_\nu(z^0))$ in ∂D to the origin, U_ν is a unitary transform sending $T_\nu(F(\varphi_\nu(z^0)))$ to the point with coordinates $(0', -\delta_\nu)$, where $\delta_\nu = dist_{Eucl}(F(\varphi_\nu(z^0)), F(\partial D \cap U))$, and Λ_ν is the anisotropic dilation preserving $\mathbb{H}_n$ given by $\Lambda_\nu(z', z_n) = (z'/(\delta_\nu)^{1/2}, z_n/\delta_\nu)$.

Claim 3 After extraction, the sequence $(A_\nu \circ F \circ \varphi_\nu)_\nu$ converges, uniformly on compact subsets of $\mathbb{C}^n$, to a biholomorphism $\Phi : D \to \mathbb{H}^n$.

The proof of Claim 3 is based on normal families arguments. One can prove that $(A_\nu \circ F \circ \varphi)_\nu$ is a normal family, due to Claim 2. Since all these maps send z^0 to $(0', -1) \in \mathbb{H}_n$, the sequence $(A_\nu \circ F \circ \varphi_\nu)_\nu$ converges, up to some extraction, to some holomorphic map $\Phi : D \to \mathbb{H}_n$. We still denote the extracted sequence $(A_\nu \circ F \circ \varphi_\nu)_\nu$. It is now easy to prove that D is a *taut* domain (see the precise definition in the next Section), implying that $((A_\nu \circ F \circ \varphi_\nu)^{-1})_\nu$ is a normal family. Since $(A_\nu \circ F \circ \varphi_\nu)^{-1}(0', -1) = z^0$ for every ν, the sequence converges to some $\Psi : \mathbb{H}_n \to D$, up to some extraction. Finally, one can prove that $\Psi = \Phi^{-1}$.

The Wong-Rosay Theorem is also true, under the same local assumptions, in complex manifolds or in Hilbert separable spaces (see [10, 17]). Strict pseudoconvexity is a complex geometric notion, it means that the boundary satisfies some complex curvature condition. That notion is probably the most natural generic curvature condition in complex geometry for open complex manifolds with boundary: the order of contact of the boundary with any non trivial holomorphic disk is equal to one. That can be generalized by quantifying the order of contact (or of tangency) of the boundary with non trivial (possibly singular) holomorphic disks. The corresponding notion was introduced by D'Angelo (see [12–14] for the definitions and for associated results): a smooth domain is of finite D'Angelo type if the supremum of the order of tangency between the boundary of the domain and non trivial analytic disks is finite. Theorem 2.2 was generalized to pseudoconvex domains of finite D'Angelo type in $\mathbb{C}^2$ by Bedford-Pinchuk [3] and Berteloot [5], and in any dimension for smooth bounded convex domains of finite type by Bedford-Pinchuk [4]. See also [29]. We point out that some dynamics study is crucial in all the generalizations mentioned above, consisting in analyzing the different parabolic vector fields whose limit (at infinity) is a pseudoconvex point of finite type. In the different approaches, the existence of such a parabolic group of automorphisms comes from the existence of a sequence of automorphisms accumulating at a boundary point of finite type. In all the previous situations, using dilations modelled on the ones presented above for strongly pseudoconvex domains, one can prove the existence of a biholomorphism between the given domain and some rigid domain $D_\varphi := \{z = (z', z_n) \in \mathbb{C}^{n-1} \times \mathbb{C}/ \ Re(z_n) + \varphi(z', \overline{z'}) < 0\}$ where φ is a pluri-subharmonic function in $\mathbb{C}^{n-1}$, without harmonic terms. The parabolic automorphism group is just the pull-back of the one-parameter group of automorphisms of D_φ given by: $(t, z) \in \mathbb{R} \times D_\varphi \mapsto (z', z_n + it)$. It is part of the work to prove that the pull-back is indeed a parabolic group. Several questions may be raised in that direction. The general question in the classification problem is probably the following:

Question 1 Is it possible to classify pseudoconvex domains in $\mathbb{C}^n$ admitting a point of finite D'Angelo type as an accumulation point?

That question is far from being understood and only very special cases, essentially presented here above, are treated. The following question is a step in the approach to a classification.

Question 2 Given a domain D in $\mathbb{C}^n$ and an accumulation point $p \in \partial D$ for $Aut(D)$, may we ensure that there exists a one parameter parabolic subgroup of $Aut(D)$ with fixed point p at infinity?

One of the main difficulties for these two questions is the lack of general dilations reflecting the geometry of the domain. The survey [6], written by Berteloot, gives a very nice presentation of the essence of the scaling method, with different applications in Complex Analysis. Question 2 is a possible (and probably essential) step in the approach to Question 1. There are few results in that direction (see [21] by Kim who proved the existence of a one parameter group for every convex domain in $\mathbb{C}^n$ admitting an accumulation point in its boundary). One essential ingredient consists in proving that certain rigid polynomial domains are taut, meaning that sequences of holomorphic maps with values in such a domain, form a normal family.

3 Kobayashi Hyperbolicity of Complex Manifolds

For the convenience of the reader, we recall the definition of tautness:

Definition 3.1 Let M be a complex manifold.

- A family of holomorphic maps from Δ to M is called *normal* if from each sequence of elements of the family we may extract a subsequence that either converges on compact subsets of Δ or is compactly divergent.

- We say that M is taut if every sequence of holomorphic maps from Δ to M forms a normal family.

The tautness notion, that appeared in Section 1, is related to the notion of Kobayashi hyperbolicity.

Definition 3.2 Let z, z' be two points in a complex manifold M and $v \in T_z M$, where $T_z M$ is the space of tangent vectors at z.

- The infinitesimal Kobayashi (pseudo)metric $k_M(z, v)$ is given by:

$$ k_M(z, v) = \inf \left\{ \alpha > 0 / \exists f \in Hol(\Delta, M), f(0) = z, f'(0) = v/\alpha \right\}. $$

- The Kobayashi (pseudo)distance is the length function defined by:

$$ K_M(z, z') = \inf \left\{ \int_0^1 k_M(\gamma(t), \gamma'(t)) dt \right\}, $$

where the infimum is taken over all piecewise C^1 curves $\gamma : [0, 1] \to M$, joining z to z'.

The Kobayashi (pseudo)metric is a Finsler (pseudo)metric that is only continuous in general. We say that M is (Kobayashi) hyperbolic if K_M is a distance (and K_M induces locally the standard topology on M). We say that M is complete hyperbolic if (M, K_M) is a complete metric space. As examples, every bounded domain in $\mathbb{C}^n$ is hyperbolic, and a complex manifold containing an entire curve, i.e. the image of $\mathbb{C}$ by a non constant holomorphic map, is not hyperbolic. One can also prove (see for instance [9, 16]) that every convex domain in $\mathbb{C}^n$ not containing a complex line is complete hyperbolic. The following strict implications are standard:

$$ M \text{ complete hyperbolic} \Rightarrow M \text{ taut} \Rightarrow M \text{ hyperbolic.} $$

The hyperbolicity of complex manifolds has been intensively studied in the context of compact manifolds. It is well known by the Brody reparametrization Lemma that a compact complex manifold is (Kobayashi) hyperbolic if and only if it is (Brody) hyperbolic, namely if it does not contain any entire curve with bounded derivative. There are many very interesting and quite long standing conjectures in that context, such as the Kobayashi conjecture:

Conjecture Every general algebraic hypersurface of dimension n and degree $d \geq 2n + 2$ in the complex projective space $\mathbb{P}^{n+1}$, is hyperbolic,

or the Green-Griffiths-Lang conjecture:

Conjecture If X is a projective variety of general type, then there exists a proper algebraic variety $Y \not\subset X$ such that every entire curve in X is contained in Y.

These two conjectures are still open in their complete forms (see the article [15] by Demailly for recent developments).

In the case of open complex manifolds, questions related to the Kobayashi hyperbolicity are of completely different nature. Notice that in case a complex manifold is hyperbolic, its automorphism group is a real Lie group of finite dimension and the question of classifying certain such manifolds is relevant; see [20] where a classification is proposed for hyperbolic manifolds with "large" automorphism groups. One may wonder which open complex manifolds are (complete) hyperbolic. For domains in $\mathbb{C}^n$, the questions are the following:

Question 3 Which unbounded domains in $\mathbb{C}^n$ are (Kobayashi) hyperbolic?

Question 4 Which geometric properties of the boundary imply that a given domain is complete hyperbolic?

Question 3 is far from being understood, even for the rigid domains D_φ introduced above. A positive answer was given by F. Berteloot in complex dimension two under some additional assumption on φ (see [5]). One may notice that the geometry of the boundary seems not to play an essential part in that question; every bounded domain in $\mathbb{C}^n$ is hyperbolic.

Recently, Harz et al. [19] defined the *core set* for unbounded strictly pseudoconvex domains. This set is the obstruction for a defining function of the domain, strictly pluri-subharmonic in a neighborhood of the boundary of the domain, to extend as a strictly pluri-subharmonic function on D. It is natural to study the possible link between the core set and the (complete) hyperbolicity of an unbounded domain. It is easy to see that a domain with a non empty core set is not hyperbolic. However, there exist domains with an empty core set that are not hyperbolic.

The geometry of the boundary of the domain in Question 4 is essential: if a domain is taut, then it is pseudoconvex. The existence of a peak holomorphic function at each boundary point is a sufficient (strong) condition that ensures the complete hyperbolicity of the domain; this is the case for bounded strictly pseudoconvex domains in $\mathbb{C}^n$, or for smooth bounded domains of finite D'Angelo type in $\mathbb{C}^2$. Notice that a strictly pseudoconvex domain, relatively compact in an almost complex manifold, is complete hyperbolic if and only if it does not contain entire curves.

4   Geometry of Complete Hyperbolic Manifolds

The geometric study of complete hyperbolic manifolds is of great interest. First, they enter the general framework of complete metric spaces. Since biholomorphisms are isometries for the Kobayashi distance, extension phenomenons for biholomorphisms are natural questions in the area. The second reason is that the Kobayashi metric being a length function, the general Hopf-Rinow Theorem implies that a complete hyperbolic manifold (M, K_M) is a geodesic space: given two points

$p, q \in M$, there exists a geodesic $\gamma : [a, b] \subset \mathbb{R} \to M$ such that $\gamma(a) = p$, $\gamma(b) = q$, and γ is an isometry from $[a, b]$ endowed with the absolute value, to M endowed with K_M. One might imagine that the complex geometry of a complete hyperbolic manifold should impact (and be impacted by) the behaviour of (real) geodesics contained in M.

4.1 Complex and Metric Geometries

As already presented in the first section of the article, the notion of finite type plays an essential role in the study of non compact complex manifolds with boundary. It is the natural context for subellipticity of the $\bar{\partial}$-Neumann operator and explains how the geometry of the boundary acts on the function theory of the domain (see [14]). This may be an obstruction to the biholomorphic equivalence between two given complex manifolds, or more generally to the existence of a Cauchy-Riemann (CR) diffeomorphism between smooth germs of submanifolds. The general question in that context is the following:

Question 5 Let M, N be two germs or real submanifolds in $\mathbb{C}^n$, with $n \geq 2$. Does there exist a smooth diffeomorphism between M and N?

The CR equivalence between CR manifolds has been studied intensively, with very striking results. The first breakthrough is due to Chern and Moser [11] who developed the theory of normal forms for Levi non degenerate hypersurfaces in $\mathbb{C}^n$. Later on, the theory was pursued further and the CR equivalence problem gained much attention in the field of Several Complex Variables. The notion of D'Angelo type being a CR invariant, it is an example of obstruction to such an existence; there may exist a CR diffeomorphism between two such smooth real hypersurfaces only if they have the same D'Angelo type.

Since the type reflects partially the local geometry of a domain near a boundary point, one might hope to get information on the type by considering (real) geodesics for the Kobayashi distance. Going back to the Poincaré Theorem presented at the beginning of the article, one may obtain the non equivalence between $\mathbb{B}^2$ and $\Delta \times \Delta$ by purely (geo)metric considerations. Given any two points in $\mathbb{B}^2$, there is a unique geodesic joining them, whereas it is not difficult to construct several geodesics joining certain points in $\Delta \times \Delta$; both $\mathbb{B}^2$ and $\Delta \times \Delta$ are complete hyperbolic manifolds, but with quite different metric properties.

One can push further the study of geodesics. Given three distinct points in a geodesic metric space X, a triangle is the union of three geodesics joining these points. Notice that there may be more than one triangle corresponding to the same three points. Let $\delta > 0$. A geodesic metric space (X, d) is called δ-thin if every side of each triangle is contained in a δ-neighborhood of the union of the two other sides of the triangle. Finally, a geodesic metric space is *Gromov hyperbolic* if it is δ-thin for some $\delta > 0$. The unit ball is Gromov hyperbolic whereas the bidisc is not. Every biholomorphism being an isometry for the corresponding Kobayashi distances, the Gromov hyperbolicity is a complex invariant. More precisely, if f is a biholomorphism between two complete hyperbolic complex manifolds M and N, then (M, K_M) is Gromov hyperbolic if and only if (N, K_N) is Gromov hyperbolic.

It was proved by Balogh and Bonk [1] that a bounded strictly pseudoconvex domain in $\mathbb{C}^n$ is Gromov hyperbolic, whereas the product of complex manifolds is never Gromov hyperbolic. It was proved in [18] that a smooth bounded convex domain containing a non trivial analytic disk in its boundary is not Gromov hyperbolic, and conjectured that the true obstruction might be that such a domain is of infinite type. That was proved by Zimmer in [28]:

Theorem 4.1 *Let D be a smooth C^∞ bounded, convex domain in $\mathbb{C}^n$. Then (D, K_D) is Gromov hyperbolic if and only if D is of finite D'Angelo type.*

That very nice result presents a bridge between a pure statement from metric geometry, namely that the space is Gromov hyperbolic, and a pure statement of complex geometry, namely that the domain is of finite type. Notice that the regularity assumption on the boundary of the domain is inherent to the consideration of domains of finite D'Angelo type; the only embarrassing property is the convexity assumption, being a pure Euclidean statement. Some regularity assumption on the boundary of the domain is necessary, see [30]. See also [23] for sufficient conditions in the non-smooth case. One could wonder if Theorem 4.1 still holds for pseudoconvex domains:

Question 6 Let D be a smooth, bounded, pseudoconvex domain in $\mathbb{C}^n$. Then D is of finite D'Angelo type if and only if (D, K_D) is Gromov hyperbolic.

The fact that such domains are complete hyperbolic is not trivial in any dimension. In complex dimension 2, it is a direct consequence of the existence of holomorphic peak functions at each boundary point (see [2]); the existence of (local) peak holomorphic functions in any dimension is still open. Question 6 is already interesting in complex dimension 2, the proof presented in [28] relying in an essential way on the geometric properties of convex domains. The issue here is for sure to prove that a smooth, bounded, pseudoconvex domain in $\mathbb{C}^2$, with (at least) one point of infinite D'Angelo type, is not Gromov hyperbolic. All the ingredients for the opposite implication are indeed available.

4.2 Topology of Complete Hyperbolic Manifolds

As we already observed, there is no hope to classify all complete hyperbolic manifolds and one seeks either to reduce the family of manifolds under consideration, or to find invariants that provide obstructions to the equivalence. The Gromov hyperbolicity is such an isometric invariant that may be an obstruction to the existence of an isometry between two given geodesic complete (called proper) metric spaces. The idea of the following is to define a boundary for complete hyperbolic manifolds in terms of horospheres, and to endow it with an associated topology. The different properties of the corresponding topological spaces may be obstructions to the existence of isometries between two given such manifolds. Although all the notions are defined for complete hyperbolic manifolds, and hence for the Kobayashi distance, they are analogous in the general framework of complete metric spaces. In the context of domains in $\mathbb{C}^n$, that theory provided new tools to study the question of extension of biholomorphisms between non necessarily smooth domains, see [8]. The following considerations are a part of a joint work with Filippo Bracci [7]. For sake of completeness, we present the construction in details.

Let us fix a point x in a complete hyperbolic manifold (M, K_M). Then by definition, for every proper sequence $\{u_\nu\}$ of points in M, we have $\lim_{\nu \to \infty} K_M(x, u_\nu) = \infty$. Given such a sequence, one may define the sequence horosphere: for every $r > 0$,

$$E_x(u_\nu, r) = \left\{ z \in M / \limsup_{\nu \to \infty} [K_M(z, u_\nu) - K_M(x, u_\nu)] < \frac{1}{2} \log r \right\}.$$

Then a sequence $\{u_\nu\}$ is called admissible if $E_x(u_\nu, r) \neq \emptyset$ for every $r > 0$. It is clear that only small values of r are interesting here. Two admissible sequences $\{u_\nu\}$ and $\{v_\nu\}$ are equivalent if, for every $r > 0$, there exist $r', r'' > 0$ such that $E_x(u_\nu, r') \subset E_x(v_\nu, r)$ and $E_x(v_\nu, r'') \subset E_x(u_\nu, r)$. This is indeed an equivalence relation and we define the horosphere boundary $\partial_H M$ as the set of equivalence classes for that relation. It is important to notice that the notions of admissible sequence, the definition of equivalent admissible sequences and hence the definition of horosphere boundary, do not depend on the fixed point x; we would obtain the exact same horosphere boundary replacing x by any other point in M. Moreover, it was pointed out by A. Zimmer that $\partial_H M \neq \emptyset$ for every complete hyperbolic manifold M. As a set, $\partial_H M$ does not seem to carry much information. The essential part of the information comes from the topology we endow $M \cup \partial_H M$ with, constructed by defining convergence of sequences. The following definitions are taken from [7]:

Definition 4.2 A sequence $\{y_m\} \subset \partial_H M$ converges to $y \in \partial_H M$ if there exist admissible sequences $\{u_\nu^m\}_{\nu \in \mathbb{N}}$ with $[\{u_\nu^m\}] = y_m$ and an admissible sequence $\{u_\nu\}$ with $[\{u_\nu\}] = y$ with the property that for every $R > 0$ there exists $m_R \in \mathbb{N}$ such that: $E_x(\{u_\nu^m\}, \bar{R}) \cap E_x(\{u_\nu\}, R) \neq \emptyset$, $\forall m \geq m_R$.

Definition 4.3 A sequence $\{z_m\} \subset M$ converges to $y \in \partial_H M$ if there exist $\{u_\nu\}, \{v_\nu^j\}_{j \in \mathbb{N}} \subset \Lambda_M$ with $[\{u_\nu\}] = y$ and with the property that for all $R > 0$, there exists $m_R \in \mathbb{N}$ such that $z_m \in E_x(\{v_\nu^m\}, R)$, and $E_x(\{v_\nu^m\}, R) \cap E_x(\{v_\nu\}, R) \neq \emptyset$, for all $m \geq m_R$.

Finally, we define closed sets using the convergence of sequences and open sets as complements of closed sets. The corresponding topology on $M \cup \partial_H M$ is called the *Horosphere topology* and is denoted by $\tau_H M$.

The first observation is that the horosphere topology does not depend on the choice of a base point. The second observation is that defining a topology using sequences might be a weak approach, since one may not ensure that this will be Hausdorff. In our situation, that may become an obstruction for the equivalence problem. As an example, we have the following:

Example 4.4

(i) $(\mathbb{B}^2 \cup \partial_H \mathbb{B}^2, \tau_H \mathbb{B}^2)$ is homeomorphic to $\overline{\mathbb{B}^2}$,
(ii) The restriction of $\tau_H(\Delta \times \Delta)$ to $\partial_H(\Delta \times \Delta)$ is trivial.

Question 7 If (M, K_M) is Gromov hyperbolic, is $(M \cup \partial_H M, \tau_H M)$ a Hausdorff space?

We proved in [7] that if D is a bounded, strictly pseudoconvex domain in $\mathbb{C}^n$, with boundary of class $\mathcal{C}^3$, then $(D \cup \partial_H D, \tau_H D)$ is homeomorphic to $\overline{D}$. In particular, according to [1], it is also homeomorphic to the Gromov closure of D, endowed with its topology.

Question 8 If (M, K_M) is Gromov hyperbolic, is $(M \cup \partial_H M, \tau_H M)$ homeomorphic to the closure of M endowed with the Gromov topology?

Acknowledgements The author would like to thank the organizers of the Workshop for the warm reception and the fruitful atmosphere. The author is partially supported by ERC ALKAGE.

References

1. Balogh, Z., Bonk, M.: Gromov hyperbolicity and the Kobayashi metric on strictly pseudoconvex domains. Comment. Math. Helv. **75**, 504–533 (2000)
2. Bedford, E., Fornaess, J.E.: A construction of peak functions on weakly pseudoconvex domains. Ann. Math. **107**, 555–568 (1978)
3. Bedford, E., Pinchuk, S.: Domains in $\mathbb{C}^2$ with noncompact holomorphic automorphism groups. Math. Sb. **135**, 147–157 (1988)
4. Bedford, E., Pinchuk, S.: Convex domains with noncompact automorphism groups. Math. Sb. **185**, 3–26 (1994)
5. Berteloot, F.: Characterization of models in $\mathbb{C}^2$ by their automorphism groups. Int. J. Math. **5**, 619–634 (1994)
6. Berteloot, F.: Méthodes de changement d'échelles en analyse complexe. Ann. Fac. Sci. Toulouse **15**, 427–483(2006)
7. Bracci, F., Gaussier, H.: Horosphere topology. ArXiv:1605.04119
8. Bracci, F., Gaussier, H.: A proof of the Muir-Suffridge conjecture for convex maps of the unit ball in $\mathbb{C}^n$. Math. Ann. doi:10.1007/s00208-017-1581-8
9. Bracci, F., Saracco, A.: Hyperbolicity in unbounded convex domains. Forum Math. **21**, 815–825 (2009)
10. Byun, J., Gaussier, H., Lee, K.H.: On the automorphism group of strongly pseudoconvex domains in almost complex manifolds. Ann. Inst. Fourier **59**, 291–310 (2009)
11. Chern, S.S., Moser, J.: Real hypersurfaces in complex manifolds. Acta math. **133**, 219–271 (1974)
12. D'Angelo, J.P.: Finite type conditions for real hypersurfaces. J. Differ. Geom. **14**, 59–66 (1979)

13. D'Angelo, J.P.: Real hypersurfaces, orders of contact, and applications. Ann. Math. **115**, 615–637 (1982)
14. D'Angelo, J.P., Kohn, J.: Subelliptic estimates and finite type. In: Several Complex Variables. MSRI Publications, vol. 37, pp. 199–232. Cambridge University Press, Cambridge (1999)
15. Demailly, J.P.: Recent progress towards the Kobayashi and Green-Griffiths-Lang conjectures (expanded version of talks given at the 16th Takagi lectures). https://www-fourier.ujf-grenoble.fr/~demailly/manuscripts/takagi16_jpd.pdf
16. Gaussier, H.: Tautness and complete hyperbolicity of domains in $\mathbb{C}^n$. Proc. Am. Math. Soc. **127**, 105–116 (1999)
17. Gaussier, H., Kim, K.T., Krantz, S.: A note on the Wong-Rosay Theorem in complex manifolds. Complex Var. Theory Appl. **47**, 761–768 (2002)
18. Gaussier, H., Seshadri, H.: On the Gromov hyperbolicity of convex domains in $\mathbb{C}^n$. To appear in CMFT. ArXiv:1312.0368
19. Harz, T., Shcherbina, N., Tomassini, G.: On defining functions and cores for unbounded domains. I. Math. Z. **286**, 987–1002 (2017)
20. Isaev, A.: Lectures on the Automorphism Groups of Kobayashi-Hyperbolic Manifolds. Lecture Notes in Mathematics, vol. 902.
21. Kim, K.T.: On the automorphism groups of convex domains in $\mathbb{C}^n$. Adv. Geom. **4**, 33–40 (2004)
22. Kobayashi, S.: Hyperbolic Complex Spaces. Grundlehren der mathematischen Wissenschaften, vol. 318. Springer, Berlin (1998)
23. Nikolov, N., Thomas, P., Trybula, M.: Gromov (non)hyperbolicity of certain domains in $\mathbb{C}^n$. Forum Math. **28**, 783–794 (2016)
24. Pinchuk, S.: The scaling method and holomorphic mappings, Several complex variables and complex geometry, Part 1 (Santa Cruz, CA, 1989), vol. 52, pp. 151–161, American Mathematical Society, Providence (1991)
25. Poincaré, H.: Les fonctions analytiques de deux variables et la représentation conforme. Rend. Circ. Mat. Palermo, pp. 185–220 (1907)
26. Rosay, J.P.: Sur une caractérisation de la boule parmi les domaines de $\mathbb{C}^n$ par son groupe d'automorphismes. Ann. Inst. Fourier **29**, 91–97 (1979)
27. Wong, B.: Characterization of the unit ball in $\mathbb{C}^n$ by its automorphism group. Invent. Math. **41**, 253–257 (1977)
28. Zimmer, A.: Gromov hyperbolicity and the Kobayashi metric on convex domains of finite type. Math. Ann. **365**, 142–198 (2016)
29. Zimmer, A.: Characterizing domains by the limit set of their automorphism group. Adv. Math. **308**, 438–482 (2017)
30. Zimmer, A.: Gromov hyperbolicity, the Kobayashi metric, and C-convex sets. Trans. Am. Math. Soc. **369**, 8437–8456 (2017)

On a Solution of a Particular Case of Aliaga-Tuneski Question

Dov Aharonov and Uri Elias

Abstract Z. Nehari found sufficient conditions implying univalence of analytic functions, expressed in terms of the Schwarzian derivative. His theorem had been used by us to generate additional families of conditions for univalence, depending on parameters.

Recently, Aliaga and Tuneski used the method of selecting parameters to find a new criterion for univalence of analytic functions. However, they presented determining specific suitable values of the parameters and their precise domain of admissibility for this family as an open question.

In what follows we obtain additional information about the exact domain of possible parameters in the family of criteria studied by Aliaga and Tuneski.

Keywords Univalent functions · Univalence criteria

Mathematics Subject Classification 30C55

1 Introduction and Preliminaries

The Schwarzian derivative $Sf = \left(f''/f'\right)' - \frac{1}{2}\left(f''/f'\right)^2$ of an analytic locally univalent function plays an important role for finding sufficient conditions for univalence. Nehari [7] found conditions implying univalence expressed in terms of the Schwarzian derivative:

D. Aharonov · U. Elias (✉)
Technion, Haifa, Israel
e-mail: dova@technion.ac.il; elias@technion.ac.il

© Springer International Publishing AG, part of Springer Nature 2017
F. Bracci (ed.), *Geometric Function Theory in Higher Dimension*,
Springer INdAM Series 26, https://doi.org/10.1007/978-3-319-73126-1_12

If $|Sf| \le 2(1 - |z|^2)^{-2}$, *then f is univalent in the unit disc* $\Delta = \{z, |z| < 1\}$. *Also if* $|Sf| \le \pi^2/2$, *the same conclusion follows.* Later Pokornyi stated without proof the condition $|Sf| \le 4(1 - |z|^2)^{-1}$. In addition Nehari extended these results and proved a more general theorem [8, 9] concerning criteria for univalence. In his theorem he also investigated the sharpness of his conditions.

Nehari's pioneering work in [7] opened a fundamental line of research. His idea was to use that if $u(z), v(z)$ are two functions such that every linear combination $c_1 u(z) + c_2 v(z)$ has at most one zero in a domain D, then their quotient $f(z) = v(z)/u(z)$ is univalent in D. Quotients of solutions are naturally related to a differential equation through the Schwarzian derivative operator due to the following property: *Suppose we are given the linear differential equation*

$$u'' + p(z)u = 0, \tag{1.1}$$

where p(z) is an analytic function in the unit disc Δ *and* $u(z), v(z)$ *are any two linearly independent solutions of (1.1). Then*

$$S(v/u)(z) = 2p(z). \tag{1.2}$$

Later Nehari made use of the Schwarzian derivative and its above properties to arrive at more sufficient conditions for univalence.

Theorem A (Nehari,[8]) *Suppose that*

(i)   p(x) is a positive and continuous even function for $-1 < x < 1$,
(ii) p(x)(1 - x^2)^2 is nonincreasing for $0 < x < 1$,
(iii) the real valued differential equation

$$y''(x) + p(x)y(x) = 0, \qquad -1 < x < 1, \tag{1.3}$$

has a solution which does not vanish in $-1 < x < 1$.

Then any analytic function f(z) in Δ *satisfying*

$$|Sf(z)| \le 2p(|z|) \tag{1.4}$$

is univalent in the unit disc Δ.

In what follows we use the term "Nehari's function" to denote a positive even function $p(x)$ such that $p(x)(1 - x^2)^2$ is nonincreasing for $0 < x < 1$. As pointed out already in [7], the functions

$$p(x) = (1 - x^2)^{-2}, \qquad p(x) = \pi^2/4, \tag{1.5}$$

and the corresponding solutions $y(x) = (1 - x^2)^{1/2}$, $y(x) = \cos(\pi x/2)$ of the respective equations (1.3) have all the needed properties to conclude the sufficient conditions for univalence in Δ.

In [2] we used the classical Theorem A of Nehari to generate additional families of conditions for univalence depending on parameters. Let $\Lambda = (\lambda_1, \lambda_2, \ldots, \lambda_n)$ be a vector of parameters and let $u = u(z, \Lambda)$ be a family of analytic functions in Δ depending on these n parameters. For this $u = u(z, \Lambda)$ we generate

$$p(z, \Lambda) = -u''(z, \Lambda)/u(z, \Lambda) \tag{1.6}$$

and the corresponding differential equation

$$u'' + p(z, \Lambda)u = 0, \qquad z \in \Delta. \tag{1.7}$$

In addition we assume that the restriction of u to the real axis, $u(x, \Lambda)$, is the solution of the real valued differential equation

$$y''(x) + p(x, \Lambda)y(x) = 0, \qquad -1 < x < 1, \tag{1.8}$$

which does not vanish in $-1 < x < 1$.

If we can find a range of parameters such that $p(x, \Lambda)(1 - x^2)^2$ is non increasing for $0 < x < 1$, we may apply Theorem A in order to find a family of univalence criteria depending on the vector Λ.

We mention a few families of parametric functions and the corresponding differential equations (1.8) which we had studied. The most simple example [1, Theorem 1] is generated by $u(x, \lambda) = (1 - x^2)(1 - \lambda x^2)$ and the corresponding Nehari function is

$$p(x, \lambda) = \frac{2(1 + \lambda) - 12\lambda x^2}{(1 - x^2)(1 - \lambda x^2)}, \qquad 3 - \sqrt{10} \le \lambda \le 1/5.$$

A two-parametric family may be generated by $u(x, \lambda, \mu) = (1 - x^2)^\lambda \cos^\mu(\frac{\pi x}{2})$ and it yields the Nehari functions

$$p(x, \lambda, \mu) = \frac{4\lambda(1 - \lambda)x^2}{(1 - x^2)^2} + \frac{2\lambda}{1 - x^2} + \frac{\mu\pi^2}{4}$$

$$+ \frac{\mu(1 - \mu)\pi^2}{4} \tan^2(\pi x/2) - 2\mu\lambda\pi \frac{x\tan(\pi x/2)}{1 - x^2},$$

with $\lambda \ge 0$, $\mu \ge 0$, $1/2 \le \lambda + \mu \le 1$, $(1 - \mu)/2 \le \lambda \le 1 - \mu$. See [2, Theorem 2]. Additional families of Nehari functions may be found in the articles [1–4].

Aliaga and Tuneski obtained analogous results with

$$u(x, a, \lambda) = \frac{1 - x^2}{a - x^2} \exp(\lambda x^2),$$

$$p(x, a, \lambda) = \frac{8(a - 1)x^2}{(1 - x^2)(a - x^2)^2} + \frac{2(a - 1)(1 + 4\lambda x^2)}{(1 - x^2)(a - x^2)} - 2\lambda(2\lambda x^2 + 1).$$

In [5, Theorem 4] they show the existence of values of λ and a such that $p(x, a, \lambda)$ is a Nehari function, but they presented determining specific suitable values for a and λ as an open question. In [10] it is shown that $-1/4 \leq \lambda \leq (3 - \sqrt{7})/4$ and sufficiently large values of a yield a Nehari function. In [6, Example 2.1] it is verified that $\lambda = -1/4$, $a = 2$ are admissible values of the parameters.

In the present work we supply a partial answer to the question of finding suitable values of the parameters:

Theorem 1 *For $\lambda = -1/4$ there exists $a^* > 1$ such that $p(x, a, \lambda)(1 - x^2)^2$ has the required monotonicity property for $a > a^*$ but monotonicity fails for $a \in (1, a^*)$. According to numerical calculations, $a^* \approx 1.86929$.*

Proof Due to the construction of $p(x)$ through $u(x)$, all we have to show is that the function $\varphi(x) := (1 - x^2)^2 p(x)$ is positive and nonincreasing for $0 < x < 1$. With $\lambda = -1/4$ this becomes

$$\varphi(x) = \frac{8(a - 1)x^2(1 - x^2)}{(a - x^2)^2} + \frac{2(a - 1)(1 - x^2)^2}{(a - x^2)} + \frac{1}{2}(-\frac{1}{2}x^2 + 1)(1 - x^2)^2$$

and substituting $x^2 = t$, $\varphi(x) = \psi(t)$, we shall show that

$$\psi(t) = \frac{8(a - 1)t(1 - t)}{(a - t)^2} + \frac{2(a - 1)(1 - t)^2}{(a - t)} + (\frac{1}{2} - \frac{1}{4}t)(1 - t)^2$$

is positive and nonincreasing for $0 < t < 1$ for suitable values of a. The condition $p(x) > 0$ follows at once. In fact we need to check only whether $(a - t)^3 \dfrac{d\psi(t)}{dt} < 0$. By putting

$$t = 1 - (1 - t), \quad a - t = (a - 1) - (1 - t),$$

we get by direct calculation

$$(a - t)^3 \frac{d\psi(t)}{dt}$$

$$= -\frac{3}{4}(1 - t)^5 - \frac{1}{2}(1 - t)^4 + (a - 1)\left(8(1 - t) - \frac{7}{2}(1 - t)^3 - \frac{9}{4}(1 - t)^4\right)$$

$$+ (a - 1)^2\left(-8 + 16(1 - t) - \frac{15}{2}(1 - t)^2 - \frac{9}{4}(1 - t)^3\right)$$

$$+ (a - 1)^3\left(-\frac{9}{2}(1 - t) - \frac{3}{4}(1 - t)^2\right).$$

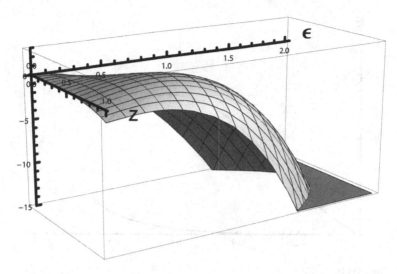

Fig. 1 The surface $q(z, \varepsilon)$, $0 \le z \le 1$, $0 \le \varepsilon \le 2$

Finally, with $z := 1 - t$, $\varepsilon := a - 1 > 0$ and $q(z, \varepsilon) := (a - t)^3 d\psi(t)/dt$, we ask whether

$$q(z, \varepsilon) = -\frac{3}{4}z^5 - \frac{1}{2}z^4 + \varepsilon\left(8z - \frac{7}{2}z^3 - \frac{9}{4}z^4\right) + \varepsilon^2\left(-8 + 16z - \frac{15}{2}z^2 - \frac{9}{4}z^3\right)$$
$$+ \varepsilon^3\left(-\frac{9}{2}z - \frac{3}{4}z^2\right)$$

is negative for $0 < z < 1$? See Fig. 1.

First we observe that for the specific value $z = \varepsilon$ one has

$$q(z, \varepsilon)\Big|_{z=\varepsilon} = 2\varepsilon^3\left(8 - 8\varepsilon - 3\varepsilon^2\right) > 0$$

for $0 < \varepsilon < \varepsilon_0 \equiv (\sqrt{40} - 4)/3 \approx 0.774852$. Hence, for $0 < \varepsilon < \varepsilon_0$ we have

$$\max_{0 \le z \le 1} q(z, \varepsilon) > 0.$$

Consequently, the function $\varphi(x)$ is not decreasing for $x \in [0, 1]$ and $p(x, a, -1/4)$ is not a Nehari function for $1 < a = 1 + \varepsilon < 1 + \varepsilon_0 \approx 1.774852$.

In order to show that $p(x, a, -1/4)$ is a Nehari function, we have to show that $q(z, \varepsilon) < 0$ for every $x \in [0, 1]$, for suitable values of ε. The value of ε where $q(z, \varepsilon)$ begins to be negative for every z may be defined as

$$\varepsilon_1 = \max\{\, \varepsilon \mid q(z, \varepsilon) < 0 \text{ for every } z \in [0, 1] \,\},$$

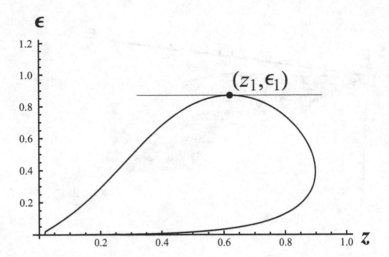

Fig. 2 The level curve $q(z, \varepsilon) = 0$

i.e., the highest point of the level curve $q(z, \varepsilon) = 0$. At this point the normal is parallel to the ε axis, so it can be found by solving the pair of equations

$$q(z, \varepsilon) = 0, \qquad \frac{\partial q}{\partial z}(z, \varepsilon) = 0.$$

By numerical calculation carried out by Mathematica, the only solution in the strip $0 < z < 1$ is approximately $(z_1, \varepsilon_1) = (0.6219054361, 0.8692914347)$. See Fig. 2.

For sufficiently large values of ε, $q(z, \varepsilon)$ is also a decreasing function of ε for every $z \in [0, 1]$. First note that the coefficient of ε^2 in $q(z, \varepsilon)$ is negative since

$$\max_{0 \le z \le 1}\left[-8 + 16z - \frac{15}{2}z^2 - \frac{9}{4}z^3\right] < -1.15 \quad \text{is obtained at } x \approx 0.787556. \text{ Next,}$$

$$\frac{dq}{d\varepsilon} = \left(8z - \frac{7}{2}z^3 - \frac{9}{4}z^4\right) + 2\varepsilon\left(-8 + 16z - \frac{15}{2}z^2 - \frac{9}{4}z^3\right) + 3\varepsilon^2\left(-\frac{9}{2}z - \frac{3}{4}z^2\right)$$

and by ignoring some negative terms,

$$< z\left(8 - \frac{27}{2}\varepsilon^2\right) < 0$$

for $\varepsilon > \sqrt{16/27} \approx 0.769800$. Consequently $(a - t)^3 d\psi(t)/dt = q(z, \varepsilon) < 0$ for every $z \in [0, 1]$ and $a = 1 + \varepsilon > 1 + \varepsilon_1 > 1.86929$. $\qquad \square$

References

1. Aharonov, D., Elias, U.: Univalence criteria depending on parameters. Anal. Math. Phys. **4**, 23–34 (2014)
2. Aharonov, D., Elias, U.: Sufficient conditions for univalence of analytic functions. 1–30, (2013). arXiv:1303.0982v1 [math.CV]
3. Aharonov, D., Elias, U.: Sufficient conditions for univalence of analytic functions. J. Anal. **22**, 1–11 (2014)
4. Aharonov, D., Elias, U.: Univalence criteria depending on parameters and applications. Contemp. Math. **667**, 15–26 (2016)
5. Aliaga, E., Tuneski, N.: On existence of sufficient condition for univalence depending on two parameters. In: Proceedings of the V Congress of Mathematicians of Macedonia, vol. 2, pp. 5–9 (2015)
6. Aliaga, E., Tuneski, N.: The Schwarzian derivative as a condition for univalence. Adv. Math. Sci. J. **5**, 111–116 (2016)
7. Nehari, Z.: The Schwarzian derivative and schlicht functions. Bull. Am. Math. Soc. **55**, 545–551 (1949)
8. Nehari, Z.: Some criteria of univalence. Proc. Am. Math. Soc. **5**, 700–704 (1954)
9. Nehari, Z.: Univalence criteria depending on the Schwarzian derivative. Ill. J. Math. **23**, 345–351 (1979)
10. Tuneski, N., Jolevska-Tuneska, B., Prangoski, B.: On existence of sharp univalence criterion using the Schwarzian derivative. C. R. Acad. Bulg. Sci. **68**, 569–576 (2015)

Loewner Chains and Extremal Problems for Mappings with A-Parametric Representation in $\mathbb{C}^n$

Ian Graham, Hidetaka Hamada, Gabriela Kohr, and Mirela Kohr

Abstract In this paper we survey various results concerning extremal problems related to Loewner chains, the Loewner differential equation, and Herglotz vector fields on the Euclidean unit ball $\mathbb{B}^n$ in $\mathbb{C}^n$. First, we survey recent results related to extremal problems for the Carathéodory families $\mathcal{M}$ and $\mathcal{N}_A$ on the Euclidean unit ball $\mathbb{B}^n$ in $\mathbb{C}^n$, where $A \in L(\mathbb{C}^n)$ with $m(A) > 0$. In the second part of this paper, we present recent results related to extremal problems for the family $S_A^0(\mathbb{B}^n)$ of normalized univalent mappings with A-parametric representation on the Euclidean unit ball $\mathbb{B}^n$ in $\mathbb{C}^n$, where $A \in L(\mathbb{C}^n)$ with $k_+(A) < 2m(A)$. In the last section we survey certain results related to extreme points and support points for a special compact subset of $S_A^0(\mathbb{B}^n)$ consisting of bounded mappings on $\mathbb{B}^n$. Particular cases, open problems, and questions will be also mentioned.

Keywords Carathéodory family · Extreme point · Herglotz vector field · Loewner chain · Loewner differential equation · Support point

2000 Mathematics Subject Classification Primary 32H02; Secondary 30C45

I. Graham
Department of Mathematics, University of Toronto, Toronto, ON, Canada
e-mail: graham@math.toronto.edu

H. Hamada
Faculty of Science and Engineering, Kyushu Sangyo University, Higashi-ku, Fukuoka, Japan
e-mail: h.hamada@ip.kyusan-u.ac.jp

G. Kohr (✉) · M. Kohr
Faculty of Mathematics and Computer Science, Babeş-Bolyai University, Cluj-Napoca, Romania
e-mail: gkohr@math.ubbcluj.ro; mkohr@math.ubbcluj.ro

© Springer International Publishing AG, part of Springer Nature 2017
F. Bracci (ed.), *Geometric Function Theory in Higher Dimension*,
Springer INdAM Series 26, https://doi.org/10.1007/978-3-319-73126-1_13

1   Introduction and Preliminaries

Let $\mathbb{C}^n$ denote the space of n complex variables $z = (z_1, \ldots, z_n)$ with the Euclidean inner product $\langle z, w \rangle = \sum_{j=1}^{n} z_j \overline{w}_j$ and the Euclidean norm $\|z\| = \langle z, z \rangle^{1/2}$. The open ball $\{z \in \mathbb{C}^n : \|z\| < \rho\}$ is denoted by $\mathbb{B}_\rho^n$ and the unit ball $\mathbb{B}_1^n$ is denoted by $\mathbb{B}^n$. The closed ball $\{z \in \mathbb{C}^n : \|z\| \leq \rho\}$ is denoted by $\overline{\mathbb{B}}_\rho^n$. In the case of one complex variable, $\mathbb{B}^1$ is denoted by $\mathbb{U}$.

Let $L(\mathbb{C}^n, \mathbb{C}^m)$ denote the space of linear operators from $\mathbb{C}^n$ into $\mathbb{C}^m$ with the standard operator norm and let I_n be the identity in $L(\mathbb{C}^n)$, where $L(\mathbb{C}^n) = L(\mathbb{C}^n, \mathbb{C}^n)$. If Ω is a domain in $\mathbb{C}^n$, let $H(\Omega)$ be the family of holomorphic mappings from Ω into $\mathbb{C}^n$ with the compact-open topology. If $f \in H(\mathbb{B}^n)$, we say that f is normalized if $f(0) = 0$ and $Df(0) = I_n$. Let $\mathcal{LS}(\mathbb{B}^n)$ be the family of normalized locally biholomorphic mappings on $\mathbb{B}^n$. Also, let $S(\mathbb{B}^n)$ be the family of normalized biholomorphic (univalent) mappings on $\mathbb{B}^n$. We denote by $S^*(\mathbb{B}^n)$ the subset of $S(\mathbb{B}^n)$ consisting of starlike mappings on $\mathbb{B}^n$.

We shall use the following notions for an operator $A \in L(\mathbb{C}^n)$ (see [35]):

$$m(A) = \min\{\Re\langle A(z), z \rangle : \|z\| = 1\}, \quad k(A) = \max\{\Re\langle A(z), z \rangle : \|z\| = 1\},$$

$$k_+(A) = \max\{\Re\lambda : \lambda \in \sigma(A)\}, \quad |V(A)| = \max\{|\langle A(z), z \rangle| : \|z\| = 1\},$$

where $\sigma(A)$ is the spectrum of A, $k_+(A)$ is the upper Lyapunov index of A, and $|V(A)|$ is the numerical radius of the operator A. Then $k_+(A) \leq k(A) \leq |V(A)| \leq \|A\|$ and it is known that $\|A\| \leq 2|V(A)|$ (see [21, Theorem 1.3.1]) and $k_+(A) = \lim_{t \to \infty} \frac{\log \|e^{tA}\|}{t}$ (see [35, p. 311]).

The authors in [10, Lemma 2.1] proved that if $A \in L(\mathbb{C}^n)$ and $u \in \mathbb{C}^n$ with $\|u\| = 1$, then the following conditions hold:

$$e^{-k(A)t} \leq \|e^{-tA}(u)\| \leq e^{-m(A)t} \quad \text{and} \quad e^{m(A)t} \leq \|e^{tA}(u)\| \leq e^{k(A)t}, \quad t \geq 0.$$

Also, $e^{tk_+(A)} \leq \|e^{tA}\|$, for all $t \geq 0$ (see e.g. [10, Remark 2.8]). These relations were useful in the study of the solutions to the generalized Loewner differential equation on $\mathbb{B}^n$, and also to extremal problems for mappings with A-parametric representation on $\mathbb{B}^n$ (see [17, 19]).

If $h \in H(\mathbb{B}^n)$, let $P_m(z) = \frac{1}{m!} D^m h(0)(z^m)$ for $m \in \mathbb{N}$, where

$$D^m h(0)(z^m) = D^m h(0)(\underbrace{z, \ldots, z}_{m-\text{times}}), \quad z \in \mathbb{C}^n,$$

and $D^m h(0)$ is the m-th Fréchet derivative of h at 0.

Harris [24, Theorem 1] proved that if P_m is a homogeneous polynomial of degree m in $\mathbb{C}^n$, then

$$\|P_m\| \leq k_m |V(P_m)|, \tag{1.1}$$

where $k_1 = e$, $k_m = m^{m/(m-1)}$ for $m \geq 2$, $\| P_m \| = \max\{\| P_m(z) \| : \|z\| = 1\}$, and

$$|V(P_m)| = \max \left\{ |\langle P_m(z), z \rangle| : \|z\| = 1 \right\}$$

is the numerical radius of P_m. Later, the constant k_1 was improved to $k_1 = 2$, in the case of $\mathbb{C}^n$ with the Euclidean norm (see [21, Theorem 1.3.1]). Various applications of Harris' estimates in geometric function theory may be found in [35] and [18].

Throughout this paper, we denote by

$$\mathcal{A} = \left\{ A \in L(\mathbb{C}^n) : k_+(A) < 2m(A) \right\}. \tag{1.2}$$

Then $I_n \in \mathcal{A}$, and if $A \in \mathcal{A}$, then $m(A) > 0$ (see [17]).

Definition 1.1 (See [39] and [20]) Let $A \in L(\mathbb{C}^n)$ with $m(A) > 0$, and let $f \in H(\mathbb{B}^n)$ be a normalized mapping. We say that f is spirallike with respect to A or A-spirallike (denoted by $f \in \widehat{S}_A(\mathbb{B}^n)$) if $f \in S(\mathbb{B}^n)$ and $f(\mathbb{B}^n)$ is a spirallike domain with respect to A, i.e. $e^{-tA}f(z) \in f(\mathbb{B}^n)$, for all $z \in \mathbb{B}^n$ and $t \geq 0$.

Recall that if $A \in L(\mathbb{C}^n)$ with $m(A) > 0$, and if $f \in \mathcal{LS}(\mathbb{B}^n)$, then $f \in \widehat{S}_A(\mathbb{B}^n)$ if and only if (see [39]; see also [20])

$$\Re \langle [Df(z)]^{-1} Af(z), z \rangle > 0, \quad z \in \mathbb{B}^n \setminus \{0\}.$$

Definition 1.2 Let $\mathcal{G}$ be a nonempty compact subset of $H(\mathbb{B}^n)$.

(i) A point $f \in \mathcal{G}$ is called an *extreme point* of $\mathcal{G}$ provided $f = tg + (1-t)h$, where $t \in (0, 1)$, $g, h \in \mathcal{G}$, implies $f = g = h$.

(ii) A point $f \in \mathcal{G}$ is called a *support point* of $\mathcal{G}$ if $\Re \Lambda(f) = \max_{g \in \mathcal{G}} \Re \Lambda(g)$ for some continuous linear functional $\Lambda : H(\mathbb{B}^n) \to \mathbb{C}$ such that $\Re \Lambda$ is nonconstant on $\mathcal{G}$.

Let $\mathrm{ex}\, \mathcal{G}$ (respectively $\mathrm{supp}\, \mathcal{G}$) be the subset of $\mathcal{G}$ consisting of extreme points of $\mathcal{G}$ (respectively support points of $\mathcal{G}$).

2 Extremal Problems for the Carathéodory Families $\mathcal{M}$ and $\mathcal{N}_A$

2.1 The Carathéodory Family $\mathcal{M}$

The following families of holomorphic mappings on $\mathbb{B}^n$ are generalizations to higher dimensions of the Carathéodory family in $\mathbb{C}^n$ (see [20, 29]):

$$\mathcal{N} = \left\{ h \in H(\mathbb{B}^n) : h(0) = 0,\ \Re \langle h(z), z \rangle > 0,\ z \in \mathbb{B}^n \setminus \{0\} \right\},$$

$$\mathcal{N}_A = \left\{ h \in \mathcal{N} : Dh(0) = A \right\}, \quad \text{where } A \in L(\mathbb{C}^n),\ m(A) > 0.$$

The family $\mathcal{N}_{I_n}$ is denoted by $\mathcal{M}$. If $n = 1$, then $f \in \mathcal{M}$ if and only if $p \in \mathcal{P}$, where $p(z) = f(z)/z, z \in \mathbb{U} \setminus \{0\}, p(0) = 1$, and

$$\mathcal{P} = \{p \in H(\mathbb{U}) : p(0) = 1, \Re p(z) > 0, z \in \mathbb{U}\}.$$

These families are useful in the study of Loewner chains and the associated Loewner differential equation on $\mathbb{B}^n$, and in various problems related to univalence on $\mathbb{B}^n$ (see [1, 3–5, 10, 12, 13, 15, 16, 19, 22, 23, 29, 31–33, 39, 41]).

In view of Harris' estimate (1.1), Graham, Hamada, and Kohr [13, Theorem 1.2] proved the following coefficient bounds for the family $\mathcal{M}$. Other coefficient bounds for various subsets of $\mathcal{M}$ may be found in [18] and [13] (see also [12] and [31]).

Theorem 2.1 *Let* $h(z) = z + \sum_{m=2}^{\infty} P_m(z^m) : \mathbb{B}^n \to \mathbb{C}^n$ *be such that* $h \in \mathcal{M}$. *Then the following relations hold:*

(i) $|V(P_m)| \leq 2, m \geq 2$. *These bounds are sharp.*
(ii) $\| P_m \| \leq 2k_m, m \geq 2$, *where* $k_m = m^{m/(m-1)}$ *for* $m \geq 2$.

Taking into account Theorem 2.1, it would be interesting to give an answer to the following question (see [13]).

Question 2.2 *Let* $n \geq 2$ *and let* $h \in \mathcal{M}$ *be such that* $h(z) = z + \sum_{m=2}^{\infty} P_m(z)$, $z \in \mathbb{B}^n$. *What are the sharp upper bounds for* $\| P_m \|, m \geq 2$?

The following definition is due to Bracci [4, Definition 1.3]. The shearing process of Bracci [4] was very useful in proving some sharp coefficient estimates for mappings in $\mathcal{M}$, and mappings with parametric representation on the unit ball $\mathbb{B}^2$ in $\mathbb{C}^2$, and to obtain bounded support points for the family $S^0(\mathbb{B}^2)$ of normalized biholomorphic mappings with parametric representation on $\mathbb{B}^2$ (see Definition 3.5).

Definition 2.3 *Let* $\rho, \sigma \in \mathbb{C} \setminus \{0\}$ *and let* $h \in H(\mathbb{B}^2)$ *be such that* $h(0) = 0$ *and*

$$h(z) = \left(\rho z_1 + q_{0,2}^1 z_2^2 + O(|z_1|^2, |z_1 z_2|, \|z\|^3), \sigma z_2 + O(\|z\|^2)\right), \quad z = (z_1, z_2) \in \mathbb{B}^2.$$

Then the shearing $h^{[c]}$ of h is given by

$$h^{[c]}(z) = \left(\rho z_1 + q_{0,2}^1 z_2^2, \sigma z_2\right), \quad z = (z_1, z_2) \in \mathbb{B}^2.$$

In connection with Question 2.2, we mention the following sharp coefficient bound for mappings in the family $\mathcal{M}$, due to Bracci (see [4, Proposition 2.1]).

Proposition 2.4 *Let* $h = (h_1, h_2) \in H(\mathbb{B}^2)$ *be such that* $h(0) = 0$ *and* $Dh(0) = I_2$. *If* $h \in \mathcal{M}$, *then* $h^{[c]} \in \mathcal{M}$ *and* $|q_{0,2}^1| \leq \frac{3\sqrt{3}}{2}$, *where* $q_{0,2}^1 = \frac{1}{2}\frac{\partial^2 h_1}{\partial z_2^2}(0)$. *This estimate is sharp.*

Next, we recall the following growth result for the family $\mathcal{M}$. The first estimate of Theorem 2.5 is due to Pfaltzgraff [29], and the upper estimate in (ii) was obtained by Graham, Hamada, and Kohr [13, Theorem 1.2]. The lower estimate in (ii) follows from (i).

Theorem 2.5 *Let $h \in \mathcal{M}$. Then the following relations hold:*

(i) $\|z\|^2 \left(\frac{1-\|z\|}{1+\|z\|} \right) \le \Re \langle h(z), z \rangle \le \|z\|^2 \left(\frac{1+\|z\|}{1-\|z\|} \right), z \in \mathbb{B}^n$.

(ii) $r\frac{1-r}{1+r} \le \|h(z)\| \le \frac{4r}{(1-r)^2}, \|z\| = r < 1$.

Recently Bracci, Elin and Shoikhet [7], obtained the following improvement of the upper growth estimate in Theorem 2.5 (ii). Note that the following result also holds in the context of complex Banach spaces.

Theorem 2.6 *Let $h \in \mathcal{M}$. Then*

$$ \|h(z)\| \le r \left[1 + 8\frac{r(1 - r\ln 2)}{(1-r)^2} \right], \quad \|z\| = r < 1. $$

In view of Theorems 2.5 (ii) and Theorem 2.6, it would be interesting to solve the following problem, which is open in higher dimensions (cf. [13]).

Problem 2.7 *Let $n \ge 2$ and $h \in \mathcal{M}$. Find the sharp upper bounds for $\|h(z)\|$ and $\|Dh(z)\|, z \in \mathbb{B}^n$.*

The following compactness result related to the Carathéodory family $\mathcal{M}$ is a direct consequence of Theorem 2.5 (ii) and the fact that the family $\mathcal{M}$ is closed (see [13, Corollary 1.3]).

Proposition 2.8 *The family $\mathcal{M}$ is compact in the topology of $H(\mathbb{B}^n)$.*

Remark 2.9 Various growth, distortion, and coefficient bounds for certain subsets of $\mathcal{M}$ on the unit ball in $\mathbb{C}^n$ and complex Banach spaces may be found in [18].

Note that recently, Muir [27] obtained an interesting integral representation formula for the family $\mathcal{M}$ in terms of the probability measures on the unit sphere $\partial \mathbb{B}^n$, which is similar to the well known Herglotz representation for the family $\mathcal{P}$. It would be interesting to see if Muir's integral representation for $\mathcal{M}$ may be used to obtain sharp coefficient bounds, growth and distortion results for $\mathcal{M}$, in dimension $n \ge 2$.

The following example provides a basic difference between the theory in the case $n = 1$ and that in higher dimensions. The fact that the mapping $h \notin \text{ex } \mathcal{M}$, for all $n \ge 2$, was proved by Voda [40, Proposition 2.3.1]. The fact that $h \in \text{supp } \mathcal{M}$ follows by an argument similar to that in the proof of [17, Example 3.10].

Example 2.10 If $n \ge 2$ and $h : \mathbb{B}^n \to \mathbb{C}^n$ is given by

$$ h(z) = \left(z_1 \frac{1 - z_1}{1 + z_1}, \dots, z_n \frac{1 - z_n}{1 + z_n} \right), \quad z = (z_1, \dots, z_n) \in \mathbb{B}^n, \tag{2.1} $$

then $h \in \text{supp } \mathcal{M} \setminus \text{ex } \mathcal{M}$.

Taking into account the fact that $\mathcal{M}$ is compact, it would be interesting to solve the following problem in the case of higher dimensions.

Problem 2.11 *Let $n \geq 2$. Characterize the families* ex $\mathcal{M}$ *and* supp $\mathcal{M}$.

2.2 The Carathéodory Family $\mathcal{N}_A$

Next, we recall the following growth and upper bounds for the numerical radius and the norm of the coefficients of mappings in $\mathcal{N}_A$, where $A \in L(\mathbb{C}^n)$ with $m(A) > 0$. The growth estimate (i) was proved by Gurganus [20], while the upper growth estimate (iv) was obtained recently in [19] (cf. [18]). The coefficient bounds (ii) and (iii) may be found in [19].

Proposition 2.12 *Let $A \in L(\mathbb{C}^n)$ be such that $m(A) > 0$. Also, let $h \in \mathcal{N}_A$. Then the following relations hold:*

(i) $m(A)\|z\|^2 \frac{1-\|z\|}{1+\|z\|} \leq \Re\langle h(z), z\rangle \leq k(A)\|z\|^2 \frac{1+\|z\|}{1-\|z\|}$, $z \in \mathbb{B}^n$.

(ii) $|V(P_m)| \leq 2k(A)$ for $m \geq 2$, where $|V(P_m)| = \max_{\|w\|=1} |\langle P_m(w), w\rangle|$.

(iii) $\| P_m \| \leq 2m^{\frac{m}{m-1}}k(A)$ for $m \geq 2$, where $\| P_m \| = \max\{\| P_m(z)\| : \|z\| = 1\}$.

(iv) $m(A)r\frac{1-r}{1+r} \leq \|h(z)\| \leq r\left[\|A\| + 8k(A)\frac{r(1-r\log 2)}{(1-r)^2}\right]$ for $\|z\| = r < 1$.

The estimates *(i), (ii) and the lower bound of (iv) are sharp if $A \in \mathcal{A}$ is a diagonal matrix with the diagonal elements $\lambda_1, \ldots, \lambda_n$ such that $0 < \lambda_1 \leq \ldots \leq \lambda_n$.*

As a consequence of Proposition 2.12 (iv) and the fact that $\mathcal{N}_A$ is a closed family, we obtain the following compactness result (see [15, Theorem 2.15]).

Corollary 2.13 *The family $\mathcal{N}_A$ is compact in the topology of $H(\mathbb{B}^n)$, for all $A \in L(\mathbb{C}^n)$ with $m(A) > 0$.*

It would be interesting to give an answer to the following question, which is natural in view of Proposition 2.12 (cf. [15]; [18]):

Question 2.14 *Let $n \geq 2$ and let $A \in L(\mathbb{C}^n)$ be such that $m(A) > 0$. Also, let $h \in \mathcal{N}_A$ be given by $h(z) = A(z) + \sum_{m=2}^{\infty} P_m(z^m)$, $z \in \mathbb{B}^n$. What are the sharp upper bounds for $\|h(z)\|$ and $\|Dh(z)\|$, $z \in \mathbb{B}^n$. What are the sharp upper bounds for $\| P_m\|$, $m \geq 2$?*

As in the case of the family $\mathcal{M}$, the following problem is of interest in $\mathbb{C}^n$, $n \geq 2$.

Problem 2.15 *Let $n \geq 2$ and let $A \in \mathcal{A}$. Characterize* ex $\mathcal{N}_A$ *and* supp $\mathcal{N}_A$.

Let $A \in L(\mathbb{C}^2)$ be given by

$$A = \begin{bmatrix} \lambda & 0 \\ 0 & 1 \end{bmatrix}, \tag{2.2}$$

for $1 \leq \lambda < 2$. Then, we obtain (see [19, Proposition 7.2]; see also [4, Proposition 2.1 and Corollary 2.2], for $A = I_2$):

Proposition 2.16 *Let* $\lambda \in [1, 2)$ *and* $A \in \mathcal{A}$ *be given by (2.2), and let* $h = (h_1, h_2) \in H(\mathbb{B}^2)$ *be such that* $h(0) = 0$ *and* $Dh(0) = A$. *If* $h \in \mathcal{N}_A$, *then* $h^{[c]} \in \mathcal{N}_A$ *and* $|q_{0,2}^1| \leq a_0$, *where*

$$a_0 = \max \left\{ a > 0 : \lambda x^2 + y^2 - axy^2 \geq 0, x, y \geq 0, x^2 + y^2 \leq 1 \right\} \qquad (2.3)$$

and $q_{0,2}^1 = \frac{1}{2} \frac{\partial^2 h_1}{\partial z_2^2}(0)$. *In addition,* $\frac{3\sqrt{3}}{2} \leq a_0 \leq \frac{(\lambda+2)\sqrt{3}}{2}$.

Note that if $\lambda = 1$, then $a_0 = \frac{3\sqrt{3}}{2}$, where a_0 is given by (2.3).

Taking into account Proposition 2.16, it is natural to ask the following question, which is open even in the case $A = I_2$ (see [5]).

Question 2.17 *Let* $\lambda \in [1, 2)$ *and let* $A \in \mathcal{A}$ *be given by (2.2). Also, let* $h = (h_1, h_2) \in H(\mathbb{B}^2)$ *be such that*

$$h(z) = \left(\lambda z_1 + \sum_{\alpha \in \mathbb{N}^2, |\alpha| \geq 2} q_\alpha^1 z^\alpha, z_2 + \sum_{\alpha \in \mathbb{N}^2, |\alpha| \geq 2} q_\alpha^2 z^\alpha \right), \quad z = (z_1, z_2) \in \mathbb{B}^2.$$

If $h \in \mathcal{N}_A$, $k \in \mathbb{N}$, $k \geq 2$, *and* $j = 1, 2$, *then what are the sharp upper bounds for* $|q_{k,0}^j|$ *and* $|q_{0,k}^j|$?

3 Extremal Problems for the Family $S_A^0(\mathbb{B}^n)$

3.1 Herglotz Vector Fields, Loewner Chains, and the Family $S_A^0(\mathbb{B}^n)$

In this section we recall certain recent results related to extremal problems for a compact subset of $S(\mathbb{B}^n)$ which consists of mappings with A-parametric representation, where $A \in \mathcal{A}$. To this end, the following notions are useful (cf. [6, 10]; see also [15] and [17]).

Definition 3.1 ([15]; cf. [29])

(i) Let $f, g \in H(\mathbb{B}^n)$. We say that f is subordinate to g ($f \prec g$) if there exists a Schwarz mapping v (i.e. $v \in H(\mathbb{B}^n)$ and $\|v(z)\| \leq \|z\|$, $z \in \mathbb{B}^n$) such that $f = g \circ v$.

(ii) A mapping $f : \mathbb{B}^n \times [0, \infty) \to \mathbb{C}^n$ is called a univalent subordination chain if $f(\cdot, t)$ is univalent on $\mathbb{B}^n$, $f(0, t) = 0$ for $t \geq 0$, and $f(\cdot, s) \prec f(\cdot, t)$, $0 \leq s \leq t < \infty$. A univalent subordination chain $f(z, t)$ is said to be A-normalized (or an A-Loewner chain) if $Df(0, t) = e^{tA}$ for $t \geq 0$, where $A \in L(\mathbb{C}^n)$ with

$m(A) > 0$. We say that $f(z, t)$ is a Loewner chain (or a normalized univalent subordination chain) if $f(z, t)$ is I_n-normalized.

(iii) Let $f(z, t)$ be an A-normalized univalent subordination chain. Then there exists a unique Schwarz mapping $v = v(\cdot, s, t)$, called the transition mapping associated with $f(z, t)$, such that $Dv(0, s, t) = e^{(s-t)A}$ and $f(\cdot, s) = f(v(\cdot, s, t), t)$ for $t \geq s \geq 0$.

Example 3.2 Let $A \in L(\mathbb{C}^n)$ with $m(A) > 0$. It is well known that if $f \in H(\mathbb{B}^n)$, then $f \in \widehat{S}_A(\mathbb{B}^n)$ if and only if $f(z, t) = e^{tA}f(z)$ is an A-normalized univalent subordination chain (see [15]). In particular, $f \in S^*(\mathbb{B}^n)$ if and only if $f(z, t) = e^t f(z)$ is a Loewner chain (see [39]).

Definition 3.3 ([2, 6, 10]) Let $h : \mathbb{B}^n \times [0, \infty) \to \mathbb{C}^n$. We say that h is a Herglotz vector field (or a generating vector field) if the following conditions hold:

(i) $h(\cdot, t) \in \mathcal{N}$, for a.e. $t \geq 0$;
(ii) $h(z, \cdot)$ is measurable on $[0, \infty)$, for all $z \in \mathbb{B}^n$.

Remark 3.4

(i) Let $A \in \mathcal{A}$ and let $f = f(z, t) : \mathbb{B}^n \times [0, \infty) \to \mathbb{C}^n$ be an A-normalized univalent subordination chain. Then there exists a Herglotz vector field $h(z, t) = Az + \cdots$, $z \in \mathbb{B}^n$, $t \geq 0$, such that f satisfies the generalized Loewner differential equation (see [15]; cf. [13, Theorem 1.10])

$$\frac{\partial f}{\partial t}(z, t) = Df(z, t)h(z, t), \quad \text{a.e.} \quad t \geq 0, \quad \forall z \in \mathbb{B}^n. \tag{3.1}$$

(ii) Conversely, if $A \in \mathcal{A}$ and $h(z, t) = Az + \cdots$ is a Herglotz vector field, then every univalent solution $f(z, t) = e^{tA}z + \cdots$ of the generalized Loewner differential equation (3.1) is an A-normalized univalent subordination chain (see [10, Theorem 3.1] and [3]; cf. [13]).

(iii) Under the above conditions, the generalized Loewner differential equation (3.1) has a unique univalent solution $f(z, t) = e^{tA}z + \cdots$ such that $\{e^{-tA}f(\cdot, t)\}_{t \geq 0}$ is a normal family on $\mathbb{B}^n$. Moreover, for every $s \geq 0$, the following limit

$$f(\cdot, s) = \lim_{t \to \infty} e^{tA}v(\cdot, s, t) \tag{3.2}$$

holds locally uniformly on $\mathbb{B}^n$, where $v(z, s, t)$ is the transition mapping associated with $f(z, t)$. Also, $\bigcup_{t \geq 0} f(\mathbb{B}^n, t) = \mathbb{C}^n$. If $g(z, t) = e^{tA}z + \cdots$ is another univalent solution of (3.1), then $g(\cdot, t) = \Phi(f(\cdot, t))$, $t \geq 0$, where Φ is a normalized univalent mapping on $\mathbb{C}^n$ (see [10] and [14]; see also [3] and [12, Chapter 8]).

Note that the univalent mapping $f(z, t)$ given by (3.2) is called the canonical solution of the generalized Loewner differential equation (3.1) (see [10]; cf. [14]).

Definition 3.5 ([15]; cf. [13] and [31], for $A = I_n$) Let $A \in L(\mathbb{C}^n)$ with $m(A) > 0$. Also, let $f \in H(\mathbb{B}^n)$ be a normalized mapping. We say that f has A-parametric representation if there exists a Herglotz vector field $h(z,t) = Az + \cdots$ such that

$$f(z) = \lim_{t \to \infty} e^{tA} v(z,t)$$

locally uniformly on $\mathbb{B}^n$, where $v = v(z,t)$ is the unique locally absolutely continuous solution on $[0, \infty)$ of the initial value problem

$$\frac{\partial v}{\partial t} = -h(v,t) \quad \text{a.e.} \quad t \in [0, \infty), \quad v(z,0) = z, \quad \forall z \in \mathbb{B}^n.$$

If $A = I_n$ and f has I_n-parametric representation, then f has parametric representation in the usual sense (see [13]; cf. [31]).

We denote by $S_A^0(\mathbb{B}^n)$ (respectively $S^0(\mathbb{B}^n)$) the family of mappings which have A-parametric representation (respectively the usual parametric representation).

The following result gives a useful characterization of the family $S_A^0(\mathbb{B}^n)$ in terms of A-normalized univalent subordination chains on $\mathbb{B}^n$ (see [15, Corollary 2.9] and [10]; cf. [13] and [31, 32], in the case $A = I_n$).

Theorem 3.6 *Let $A \in \mathcal{A}$ and let $f \in H(\mathbb{B}^n)$ be a normalized mapping. Then $f \in S_A^0(\mathbb{B}^n)$ if and only if there exists an A-normalized univalent subordination chain $f(z,t)$ such that $f = f(\cdot, 0)$ and $\{e^{-tA} f(\cdot, t)\}_{t \geq 0}$ is a normal family on $\mathbb{B}^n$.*

Taking into account Theorem 3.6, the authors in [15, Theorem 2.15] proved the following compactness result for the family $S_A^0(\mathbb{B}^n)$.

Theorem 3.7 *If $A \in \mathcal{A}$, then $S_A^0(\mathbb{B}^n)$ is a compact subset of $H(\mathbb{B}^n)$.*

Remark 3.8

(i) If $n = 1$, $S_a^0(\mathbb{U}) = S$, for every $a \in \mathbb{C}$, $\Re a > 0$ (see [30]). However, if $n \geq 2$ and $A \in \mathcal{A}$, then $S_A^0(\mathbb{B}^n) \subsetneqq S(\mathbb{B}^n)$ (see [15]; cf. [13]). In particular, $S^0(\mathbb{B}^n) \subsetneqq S(\mathbb{B}^n)$, for $n \geq 2$ (see [13]). In fact, $S(\mathbb{B}^n)$ is not compact in dimension $n \geq 2$, while $S_A^0(\mathbb{B}^n)$ is a compact subset of $H(\mathbb{B}^n)$, for all $A \in \mathcal{A}$.

(ii) Recently, the authors in [19], proved that if $\lambda \in (1, 2)$ and $A \in L(\mathbb{C}^2)$ is given by (2.2), then $S_A^0(\mathbb{B}^2) \neq S^0(\mathbb{B}^2)$. This is one of the motivations for the study of the family $S_A^0(\mathbb{B}^n)$, for $A \in \mathcal{A}$ and $n \geq 2$.

(iii) Let $A \in L(\mathbb{C}^n)$ be such that $A + A^* = 2aI_n$, for some $a > 0$, where A^* is the adjoint operator of A. Then $S_A^0(\mathbb{B}^n) = S^0(\mathbb{B}^n)$ ([15, Theorem 3.12, Remark 3.13]).

(iv) Let $A \in \mathcal{A}$. Then it is easily seen that $\widehat{S}_A(\mathbb{B}^n) \subseteq S_A^0(\mathbb{B}^n)$, by Theorem 3.6 (see [15]). In particular, $S^*(\mathbb{B}^n) \subseteq S^0(\mathbb{B}^n)$ (see [13], [31]).

(v) Let $A \in L(\mathbb{C}^n)$ with $m(A) > 0$. Voda [41, Theorem 3.1] proved that the family $\widehat{S}_A(\mathbb{B}^n)$ is compact if and only if A is nonresonant. In particular, if $A \in \mathcal{A}$, then $\widehat{S}_A(\mathbb{B}^n)$ is compact.

3.2 Extreme Points and Support Points for $S_A^0(\mathbb{B}^n)$

Next, we refer to certain extremal results related to the family $S_A^0(\mathbb{B}^n)$, where $A \in \mathcal{A}$ and $n \geq 2$. In particular, these results apply for the family $S^0(\mathbb{B}^n)$.

The following result was obtained by Pell [28] and Kirwan [26], in the case $n = 1$. The statement (i) of Theorem 3.9 was obtained in [17, Theorem 3.1] (see [16, Theorem 2.1], in the case $A = I_n$). The second statement (ii) was first proved by Schleissinger [38, Theorem 1.1], in the case $A = I_n$, by an interesting argument based on the notion of Runge pairs in $\mathbb{C}^n$. His result provides an affirmative answer to [16, Conjecture 2.6]. The statement (ii) of Theorem 3.9 is a generalization of [38, Theorem 1.1], in the case of the family $S_A^0(\mathbb{B}^n)$ for $A \in \mathcal{A}$, and was obtained in [17, Theorem 3.5].

Theorem 3.9 *Let $A \in \mathcal{A}$. Also let $f \in S_A^0(\mathbb{B}^n)$. Also, let $f(z, t)$ be an A-normalized univalent subordination chain such that $f = f(\cdot, 0)$ and $\{e^{-tA}f(\cdot, t)\}_{t \geq 0}$ is a normal family on $\mathbb{B}^n$. Then the following statements hold:*

(i) *If $f \in \mathrm{ex}\, S_A^0(\mathbb{B}^n)$, then $e^{-tA}f(\cdot, t) \in \mathrm{ex}\, S_A^0(\mathbb{B}^n)$, for all $t \geq 0$.*
(ii) *If $f \in \mathrm{supp}\, S_A^0(\mathbb{B}^n)$, then $e^{-tA}f(\cdot, t) \in \mathrm{supp}\, S_A^0(\mathbb{B}^n)$, for all $t \geq 0$.*

It is well known that in the case of one complex variable, every support point (respectively extreme point) of S is an unbounded mapping, which maps the unit disc $\mathbb{U}$ onto the complement of a Jordan arc with increasing modulus at ∞ (see e.g. [9, Chapter 9], [30, Chapter 7]). The authors [17, Conjecture 3.6] conjectured that no bounded mapping in $S_A^0(\mathbb{B}^n)$ is a support/extreme point for the $S_A^0(\mathbb{B}^n)$, for $A \in \mathcal{A}$ and $n \geq 2$. However, quite surprisingly is the following recent result of Bracci [4, Theorem 1.2], which shows that in higher dimensions there exist bounded support points of $S^0(\mathbb{B}^n)$. The proof of this result is based on an interesting construction due to Bracci, called "shearing process" (see [4, Theorem 1.4]).

Theorem 3.10 *Let $f = (f_1, f_2) \in S^0(\mathbb{B}^2)$ be given by*

$$f(z) = \left(z_1 + \sum_{\alpha \in \mathbb{N}^2, |\alpha| \geq 2} a_\alpha^1 z^\alpha, z_2 + \sum_{\alpha \in \mathbb{N}^2, |\alpha| \geq 2} a_\alpha^2 z^\alpha\right), \quad z = (z_1, z_2) \in \mathbb{B}^2. \quad (3.3)$$

Then $|a_{0,2}^1| \leq \frac{3\sqrt{3}}{2}$. This estimate is sharp. Moreover, if $F : \mathbb{B}^2 \to \mathbb{C}^2$ is given by

$$F(z) = \left(z_1 + \frac{3\sqrt{3}}{2} z_2^2, z_2\right), \quad z = (z_1, z_2) \in \mathbb{B}^2,$$

then F is a bounded support point of $S^0(\mathbb{B}^2)$.

Next, we present the following extension of the above result to the case of the family $S_A^0(\mathbb{B}^2)$, where $A \in \mathcal{A}$ is the matrix given by (2.2). Note that Theorem 3.11 was obtained in [19, Theorem 7.6], by using arguments similar to those in the proof of [4, Theorem 1.2]. The result contained in Theorem 3.11 shows that if A is the matrix given by (2.2), then there exist bounded support points for $S_A^0(\mathbb{B}^2)$.

Theorem 3.11 *Let $A \in \mathcal{A}$ be given by (2.2), where $\lambda \in [1, 2)$, and let $f \in S_A^0(\mathbb{B}^2)$ be given by (3.3). Also, let a_0 be given by (2.3). Then $|a_{0,2}^1| \leq a_0/(2 - \lambda)$, where $a_{0,2} = \frac{1}{2}\frac{\partial^2 f_1}{\partial z_2^2}(0)$. This estimate is sharp. Moreover, if $F : \mathbb{B}^2 \to \mathbb{C}^2$ is given by*

$$F(z) = \left(z_1 + \frac{a_0}{2 - \lambda}z_2^2, z_2 \right), \quad z = (z_1, z_2) \in \mathbb{B}^2, \tag{3.4}$$

then F is a bounded support point of $S_A^0(\mathbb{B}^2)$.

The following questions are natural (see [8] and [19]; cf. [4] and [5], for $A = I_2$).

Question 3.12 *Let $A \in \mathcal{A}$ be given by (2.2), where $\lambda \in [1, 2)$, and let $F : \mathbb{B}^2 \to \mathbb{C}^2$ be given by (3.4). Then is it true that $F \in \mathrm{ex}\, S_A^0(\mathbb{B}^2)$?*

Problem 3.13 *Let $A \in \mathcal{A}$ and $n \geq 2$. Characterize all bounded support/extreme points of the family $S_A^0(\mathbb{B}^n)$.*

Question 3.14 (cf. [5, Question 3.4], for $A = I_n$) *Let $n \geq 2$. Also, let $A \in \mathcal{A}$. Characterize the support/extreme points of the compact family $\widehat{S}_A(\mathbb{B}^n)$.*
Let $f \in \widehat{S}_A(\mathbb{B}^n)$ and let $h \in \mathcal{N}_A$ be given by $h(z) = [Df(z)]^{-1}Af(z)$, $z \in \mathbb{B}^n$. Then is it true that

(i) $f \in \mathrm{ex}\,\widehat{S}_A(\mathbb{B}^n)$ if and only if $h \in \mathrm{ex}\,\mathcal{N}_A$?
(ii) $f \in \mathrm{supp}\,\widehat{S}_A(\mathbb{B}^n)$ if and only if $h \in \mathrm{supp}\,\mathcal{N}_A$?

In the case of one complex variable, the Koebe function is a support/extreme point for the family S. However, in higher dimensions, the Koebe mapping is a support point for the family $S^0(\mathbb{B}^n)$, but is not an extreme point for $S^0(\mathbb{B}^n)$ (see [17]). The next example was obtained in [17, Remark 3.11]. Note that Voda [40, Proposition 2.3.2] proved that if F is given by (3.5), then $F \notin \mathrm{ex}S^*(\mathbb{B}^n)$, for $n \geq 2$.

Example 3.15 (See [17]) Let $F : \mathbb{B}^n \to \mathbb{C}^n$ be the Koebe mapping given by

$$F(z) = \left(\frac{z_1}{(1 - z_1)^2}, \ldots, \frac{z_n}{(1 - z_n)^2} \right), \quad z = (z_1, \ldots, z_n) \in \mathbb{B}^n. \tag{3.5}$$

Then $f \in \mathrm{supp}\, S^0(\mathbb{B}^n) \setminus \mathrm{ex}\, S^0(\mathbb{B}^n)$ for $n \geq 2$.

Remark 3.16 If $f \in S^0(\mathbb{B}^n)$, then $|V(P_2)| \leq 2$, where $P_2(z) = \frac{1}{2}D^2f(0)(z^2)$, $z \in \mathbb{C}^n$ (see [13, Theorem 2.14]). This bound is sharp, and equality holds in the case of the Koebe mapping F given by (3.5). However, it is not known the sharp upper bound of $\| P_2 \|$, for $f \in S^0(\mathbb{B}^n)$, $n \geq 2$ (see [13]). Note that in contrast to the case $n = 1$, there exist mappings $f \in S^0(\mathbb{B}^n)$ $(n \geq 2)$ with $\| P_2 \| > 2$ (see [13, Remark 2.18]).

Note that the Koebe mapping F given by (3.5) also provides sharpness in the growth result for the family $S^0(\mathbb{B}^n)$ (see [13, Corollary 2.4]; cf. [31]):

$$\frac{\|z\|}{(1+\|z\|)^2} \leq \|f(z)\| \leq \frac{\|z\|}{(1-\|z\|)^2}, \quad \forall f \in S^0(\mathbb{B}^n), \quad z \in \mathbb{B}^n.$$

In view of the above remark, it is natural to ask the following question (cf. [15]).

Question 3.17 *Let $n \geq 2$ and let $A \in \mathcal{A}$. Find the sharp growth result for the family $S_A^0(\mathbb{B}^n)$. Also, if $f \in S_A^0(\mathbb{B}^n)$ such that $f(z) = z + \sum_{m=2}^{\infty} P_m(z)$, $z \in \mathbb{B}^n$, then what are the sharp upper bounds for $|V(P_m)|$ and $\|P_m\|$, $m \geq 2$?*

Next, we recall some recent extremal results, which provide sufficient conditions for a mapping $f \in S_A^0(\mathbb{B}^n)$ to do not belong to ex $S_A^0(\mathbb{B}^n) \cup$ supp $S_A^0(\mathbb{B}^n)$, where $A \in \mathcal{A}$. Propositions 3.18 and 3.19 were obtained by Bracci, Graham, Hamada, and Kohr (see [8, Theorem 3.1, Proposition 4.5]), in the case $A = I_n$, and in [19, Theorem 4.1, Proposition 5.1, Corollaries 5.3 and 5.5], for the family $S_A^0(\mathbb{B}^n)$, where $A \in \mathcal{A}$. The first result given in Proposition 3.18 provides a construction of a family of A-normalized univalent subordination chains for $A \in \mathcal{A}$, based on some variation of a given A-normalized univalent subordination chain, which satisfies certain natural conditions.

Proposition 3.18 *Let $A \in \mathcal{A}$ and let $f(z,t)$ be an A-normalized univalent subordination chain on $\mathbb{B}^n \times [0,\infty)$ such that $\{e^{-tA}f(\cdot,t)\}_{t\geq0}$ is a normal family on $\mathbb{B}^n$. Let $h(z,t)$ be the associated Herglotz vector field of $f(z,t)$. Assume that there exist some constants $T > 0$, $a \geq 1$, $b \geq \|A\|$ and $c \in (0, m(A)]$ such that the following conditions hold:*

 (i) $\|[Df(z,t)]^{-1}\| \leq a$, for all $z \in \mathbb{B}^n$ and $t \in [0,T)$.
 (ii) $\|h(z,t)\| \leq b$, a.e. $t \in [0,T)$ and for all $z \in \mathbb{B}^n$.
 (iii) $\Re\langle h(z,t), z\rangle \geq c\|z\|^2$, a.e. $t \in [0,T)$ and for all $z \in \mathbb{B}^n$.

Then there exists some $\varepsilon > 0$ such that $f(\cdot,0) \pm \varepsilon g \in S_A^0(\mathbb{B}^n)$, for all $g \in H(\mathbb{B}^n)$ such that $g(0) = 0$, $Dg(0) = \mathbf{0}_n$, $\sup_{z\in\mathbb{B}^n} \|g(z)\| \leq 1$ and $\sup_{z\in\mathbb{B}^n} \|Dg(z)\| \leq 1$. In particular, $f(\cdot,0) \in S_A^0(\mathbb{B}^n) \setminus (\text{supp } S_A^0(\mathbb{B}^n) \cup \text{ex } S_A^0(\mathbb{B}^n))$.

Proposition 3.19 *Let $A \in \mathcal{A}$ and let $f \in S_A^0(\mathbb{B}^n)$. Assume that*

 (i) *either there exist a mapping $g \in S_A^0(\mathbb{B}^n)$ and $r \in (0,1)$ such that $f(z) = \frac{1}{r}g(rz)$, $z \in \mathbb{B}^n$,*
 (ii) *or $f \in \widehat{S}_A(\mathbb{B}^n)$ is a bounded mapping on $\mathbb{B}^n$ with $\sup_{z\in\mathbb{B}^n} \|[Df(z)]^{-1}\| < \infty$, and $\Re\langle [Df(z)]^{-1}Af(z), z\rangle \geq c\|z\|^2$, $z \in \mathbb{B}^n$, for some $c \in (0, m(A))$.*

Then $f \in S_A^0(\mathbb{B}^n) \setminus (\text{supp } S_A^0(\mathbb{B}^n) \cup \text{ex } S_A^0(\mathbb{B}^n))$.

The following extremal result related to the family $S^0(\mathbb{B}^n)$ is due to Roth (see [37, Corollary 4.9 and Remark 4.10]).

Theorem 3.20 *Let $f(z,t) = e^t z + \cdots$ be a Loewner chain such that $\{e^{-t}f(\cdot,t)\}_{t\geq0}$ is a normal family on $\mathbb{B}^n$. Also, let $h(z,t)$ be the associated Herglotz vector field of*

$f(z,t)$. *If there exists a set* $E \subseteq [0,\infty)$ *of positive measure such that*

$$\inf_{z \in \mathbb{B}^n \setminus \{0\}} \Re \langle h(z,t), z/\|z\|^2 \rangle > 0, \quad \forall\, t \in E, \tag{3.6}$$

then $f_0 \notin \operatorname{supp} S^0(\mathbb{B}^n)$, *where* $f_0 = f(\cdot, 0)$.

From Theorem 3.20, we formulate the following question (cf. [37], for $A = I_n$).

Question 3.21 *Let* $A \in \mathcal{A}$ *and* $f \in S_A^0(\mathbb{B}^n)$. *Also, let* $f(z,t)$ *be an A-normalized univalent subordination chain such that* $f = f(\cdot, 0)$ *and* $\{e^{-tA} f(\cdot, t)\}_{t \geq 0}$ *is a normal family on* $\mathbb{B}^n$. *Let* $h(z,t)$ *be the associated Herglotz vector field of* $f(z,t)$. *Assume that there exists a set* $E \subseteq [0,\infty)$ *of positive measure such that the condition* (3.6) *holds. Then is it true that* $f \notin (\operatorname{supp} S_A^0(\mathbb{B}^n) \cup \operatorname{ex} S_A^0(\mathbb{B}^n))$?

4   Extremal Problems for Bounded Mappings in $S_A^0(\mathbb{B}^n)$

In this section we indicate recent progress in some extremal problems related to bounded univalent mappings on $\mathbb{B}^n$ which have A-parametric representation, where $A \in \mathcal{A}$. To this end, we make use of certain ideas and notions from control theory, to obtain properties of the time-T-reachable family. More information and ideas may be found in [16, 17], and [19].

Let $A \in \mathcal{A}$. Also, let $M \in (1, \infty)$ and

$$S_A^0(M, \mathbb{B}^n) = \left\{ f \in S_A^0(\mathbb{B}^n) : \|f(z)\| < M,\ z \in \mathbb{B}^n \right\}.$$

The family $S_{I_n}^0(M, \mathbb{B}^n)$ is denoted by $S^0(M, \mathbb{B}^n)$. It is obvious that, in the case $n = 1$, $S_a^0(M, \mathbb{U}) = S^0(M, \mathbb{U}) = S(M)$, for $a \in \mathbb{C}$, $\Re a > 0$, by Remark 3.8 (i), where $S(M)$ is the subset of S consisting of bounded functions on $\mathbb{U}$, that is

$$S(M) = \left\{ f \in S : |f(z)| < M,\ z \in \mathbb{U} \right\}.$$

Definition 4.1 (cf. [36] and [16]; see [17, 19]) Let $E \subseteq [0, \infty)$ be an interval and let $A \in \mathcal{A}$. A mapping $h : \mathbb{B}^n \times E \to \mathbb{C}^n$ is called a Carathéodory mapping on E with values in $\mathcal{N}_A$ if the following conditions hold:

(i) $h(\cdot, t) \in \mathcal{N}_A$ for $t \in E$.
(ii) $h(z, \cdot)$ is measurable on E for $z \in \mathbb{B}^n$.

Let $\mathcal{C}(E, \mathcal{N}_A)$ be the family of all Carathéodory mappings on E with values in $\mathcal{N}_A$.

Definition 4.2 (cf. [11], [36, Chapter I], for $n = 1$; [16, 17, 19]) Let $A \in \mathcal{A}$ and let $T > 0$. Also, let $h \in \mathcal{C}([0, T], \mathcal{N}_A)$, and let $v = v(z, t; h)$ be the unique Lipschitz continuous solution on $[0, T]$ of the initial value problem

$$\frac{\partial v}{\partial t}(z, t) = -h(v(z, t), t) \quad \text{a.e.} \quad t \in [0, T], \quad v(z, 0) = z, \tag{4.1}$$

for $z \in \mathbb{B}^n$, such that $v(\cdot, t; h)$ is a univalent Schwarz mapping and $Dv(0, t; h) = e^{-tA}$ for $t \in [0, T]$. Let

$$\widetilde{\mathcal{R}}_T(\mathrm{id}_{\mathbb{B}^n}, \mathcal{N}_A) = \left\{ e^{TA} v(\cdot, T; h) : h \in \mathcal{C}([0, T], \mathcal{N}_A) \right\}, \quad T \in (0, \infty),$$

and

$$\widetilde{\mathcal{R}}_\infty(\mathrm{id}_{\mathbb{B}^n}, \mathcal{N}_A) = \left\{ \lim_{t \to \infty} e^{tA} v(\cdot, t; h) : h \in \mathcal{C}([0, \infty), \mathcal{N}_A) \right\}.$$

The family $\widetilde{\mathcal{R}}_T(\mathrm{id}_{\mathbb{B}^n}, \mathcal{N}_A)$ is called *the normalized time-T-reachable family of* (4.1), where $\mathrm{id}_{\mathbb{B}^n}$ is the identity mapping on $\mathbb{B}^n$.

Remark 4.3 It is known that $\widetilde{\mathcal{R}}_\infty(\mathrm{id}_{\mathbb{U}}, \mathcal{M}) = S$ and $\widetilde{\mathcal{R}}_{\log M}(\mathrm{id}_{\mathbb{U}}, \mathcal{M}) = S(M)$ (see [11]; see also [30] and [36, Theorem 1.48]). Also, $\widetilde{\mathcal{R}}_\infty(\mathrm{id}_{\mathbb{B}^n}, \mathcal{N}_A) = S_A^0(\mathbb{B}^n)$, by the definition of both families (cf. [17]).

The following characterization of the family $\widetilde{\mathcal{R}}_{\log M}(\mathrm{id}_{\mathbb{B}^n}, \mathcal{N}_A)$ in terms of A-normalized univalent subordination chains is useful. This characterization was used in [17] and [19], to obtain growth and extremal results related to the family $\widetilde{\mathcal{R}}_{\log M}(\mathrm{id}_{\mathbb{B}^n}, \mathcal{N}_A)$, for $A \in \mathcal{A}$. Proposition 4.4 was obtained in [17, Theorem 4.5] (see [16, Theorem 3.7], in the case $A = I_n$). In view of the following result, we deduce that $\widetilde{\mathcal{R}}_{\log M}(\mathrm{id}_{\mathbb{B}^n}, \mathcal{N}_A) \subseteq S_A^0(\mathbb{B}^n, \|e^{A \log M}\|)$ (see [17]).

Proposition 4.4 *Let $A \in \mathcal{A}$. Also, let $M > 1$ and let $f \in H(\mathbb{B}^n)$ be a normalized mapping. Then $f \in \widetilde{\mathcal{R}}_{\log M}(\mathrm{id}_{\mathbb{B}^n}, \mathcal{N}_A)$ if and only if there exists an A-normalized univalent subordination chain $f(z, t)$ such that $f(\cdot, 0) = f$, $f(\cdot, \log M) = e^{A \log M} \mathrm{id}_{\mathbb{B}^n}$ and $\{e^{-tA} f(\cdot, t)\}_{t \geq 0}$ is a normal family on $\mathbb{B}^n$.*

Remark 4.5 In view of Proposition 4.4, it is easy to deduce that if $A \in \mathcal{A}$ and $\Phi \in \widehat{S}_A(\mathbb{B}^n)$, then $\Phi_A^M \in \widetilde{\mathcal{R}}_{\log M}(\mathrm{id}_{\mathbb{B}^n}, \mathcal{N}_A)$, where $\Phi_A^M(z) = e^{A \log M} \Phi^{-1}(e^{-A \log M} \Phi(z))$, $z \in \mathbb{B}^n$ (see [17, Example 4.4]). In particular, if $A = I_n$ and $\Phi \in S^*(\mathbb{B}^n)$, then $\Phi^M \in \widetilde{\mathcal{R}}_{\log M}(\mathrm{id}_{\mathbb{B}^n}, \mathcal{M})$, where $\Phi^M(z) = M\Phi^{-1}(\Phi(z)/M), z \in \mathbb{B}^n$ (see [16, Example 3.5]).

Remark 4.6 Let $A \in \mathcal{A}$. The authors in [17, Corollary 4.7] proved that the family $\widetilde{\mathcal{R}}_{\log M}(\mathrm{id}_{\mathbb{B}^n}, \mathcal{N}_A)$ is a compact subset of $H(\mathbb{B}^n)$, for all $M \in (1, \infty)$. Obviously, $\widetilde{\mathcal{R}}_\infty(\mathrm{id}_{\mathbb{B}^n}, \mathcal{N}_A)$ is also compact, since $\widetilde{\mathcal{R}}_\infty(\mathrm{id}_{\mathbb{B}^n}, \mathcal{N}_A) = S_A^0(\mathbb{B}^n)$.

Next, we present the following generalization of Pell's and Kirwan's results to the case of the family $\widetilde{\mathcal{R}}_{\log M}(\mathrm{id}_{\mathbb{B}^n}, \mathcal{N}_A)$, where $A \in \mathcal{A}$ (compare [36, Theorem 2.52], for $n = 1$). The statement (i) of Theorem 4.7 was obtained in [17, Theorem 4.8], while the statement (ii) of Theorem 4.7 was deduced in [19, Theorem 6.8] (compare [16], in the case $A = I_n$).

Theorem 4.7 *Let $A \in \mathcal{A}$ and $M > 1$. Also, let $f \in \widetilde{\mathcal{R}}_{\log M}(\mathrm{id}_{\mathbb{B}^n}, \mathcal{N}_A)$. Let $f(z,t)$ be an A-normalized univalent subordination chain such that $f = f(\cdot, 0)$, $f(\cdot, \log M) = e^{A \log M} \mathrm{id}_{\mathbb{B}^n}$ and $\{e^{-tA}f(\cdot, t)\}_{t \geq 0}$ is a normal family on $\mathbb{B}^n$. Then the following statements hold:*

   (i) *If $f \in \mathrm{ex}\, \widetilde{\mathcal{R}}_{\log M}(\mathrm{id}_{\mathbb{B}^n}, \mathcal{N}_A)$, then $e^{-tA}f(\cdot, t) \in \mathrm{ex}\, \widetilde{\mathcal{R}}_{\log M - t}(\mathrm{id}_{\mathbb{B}^n}, \mathcal{N}_A)$, for all $t \in [0, \log M)$.*

   (ii) *If $f \in \mathrm{supp}\, \widetilde{\mathcal{R}}_{\log M}(\mathrm{id}_{\mathbb{B}^n}, \mathcal{N}_A)$, then $e^{-tA}f(\cdot, t) \in \mathrm{supp}\, \widetilde{\mathcal{R}}_{\log M - t}(\mathrm{id}_{\mathbb{B}^n}, \mathcal{N}_A)$, for all $t \in [0, \log M)$.*

The next result shows that if $f \in \widetilde{\mathcal{R}}_{\log M}(\mathrm{id}_{\mathbb{B}^n}, \mathcal{N}_A)$, then f cannot be a support/extreme point for the family $S_A^0(\mathbb{B}^n)$ (see [19, Proposition 6.9]; see [8, Corollary 5.6], in the case $A = I_n$).

Proposition 4.8 *Let $A \in \mathcal{A}$ and let $M \in (1, \infty)$. Also, let $f \in \widetilde{\mathcal{R}}_{\log M}(\mathrm{id}_{\mathbb{B}^n}, \mathcal{N}_A)$. Then $f \in S_A^0(\mathbb{B}^n) \setminus (\mathrm{supp}\, S_A^0(\mathbb{B}^n) \cup \mathrm{ex}\, S_A^0(\mathbb{B}^n))$. In particular, $\mathrm{id}_{\mathbb{B}^n} \in S_A^0(\mathbb{B}^n) \setminus (\mathrm{supp}\, S_A^0(\mathbb{B}^n) \cup \mathrm{ex}\, S_A^0(\mathbb{B}^n))$.*

The following sharp coefficient bound for mappings in the reachable family $\widetilde{\mathcal{R}}_{\log M}(\mathrm{id}_{\mathbb{B}^n}, \mathcal{N}_A)$ was first obtained in [8, Theorem 5.9], in the case $A = I_2$, and then in [19, Theorem 7.11, Remark 7.12], in the case $A \in \mathcal{A}$ given by (2.2). The proof of Theorem 4.9 uses the fact that if $\Phi \in \widehat{S}_A(\mathbb{B}^n)$, then $\Phi_A^M \in \widetilde{\mathcal{R}}_{\log M}(\mathrm{id}_{\mathbb{B}^2}, \mathcal{N}_A)$, where Φ_A^M is the mapping given in Remark 4.5 (see [19]).

Theorem 4.9 *Let $A \in \mathcal{A}$ be given by (2.2), where $\lambda \in [1, 2)$. Also, let $M > 1$. If $f = (f_1, f_2) \in \widetilde{\mathcal{R}}_{\log M}(\mathrm{id}_{\mathbb{B}^2}, \mathcal{N}_A)$ is given by (3.3), then $|a_{0,2}^1| \leq \frac{a_0}{2 - \lambda}(1 - M^{\lambda - 2})$, where a_0 is given by (2.3), and $a_{0,2}^1 = \frac{1}{2}\frac{\partial^2 f_1}{\partial z_2^2}(0)$. This estimate is sharp. Equality holds in the case of the mapping $F_A^M : \mathbb{B}^2 \to \mathbb{C}^2$ given by*

$$F_A^M(z) = \left(z_1 + \frac{a_0}{2 - \lambda}(1 - M^{\lambda - 2})z_2^2, z_2 \right), \quad z = (z_1, z_2) \in \mathbb{B}^2. \tag{4.2}$$

Moreover, $F_A^M \in \mathrm{supp}\, \widetilde{\mathcal{R}}_{\log M}(\mathrm{id}_{\mathbb{B}^2}, \mathcal{N}_A) \setminus (\mathrm{supp}\, S_A^0(\mathbb{B}^2) \cup \mathrm{ex}\, S_A^0(\mathbb{B}^2))$.

In view of Theorem 4.9, it would be interesting to give an answer to the following question (see [8, Remark 5.10], for $\lambda = 1$; [19, Conjecture 7.14], for $\lambda \in (1, 2)$).

Question 4.10 *Let $A \in \mathcal{A}$ be given by (2.2), where $\lambda \in [1, 2)$. Also, let $M > 1$. Does the mapping F_A^M given by (4.2) remain an extreme point of $\widetilde{\mathcal{R}}_{\log M}(\mathrm{id}_{\mathbb{B}^2}, \mathcal{N}_A)$?*

Remark 4.11 In [16, Conjecture 3.9], it was conjectured that $\widetilde{\mathcal{R}}_{\log M}(\mathrm{id}_{\mathbb{B}^n}, \mathcal{M}) = S^0(M, \mathbb{B}^n)$, for all $n \geq 2$. Clearly, in the case $n = 1$, $\widetilde{\mathcal{R}}_{\log M}(\mathrm{id}_{\mathbb{U}}, \mathcal{M}) = S(M)$. However, quite surprisingly, the authors in [8] proved recently that if $n \geq 2$, then there exists $M > 2$ such that $\widetilde{\mathcal{R}}_{\log M}(\mathrm{id}_{\mathbb{B}^n}, \mathcal{M}) \subsetneqq S^0(M, \mathbb{B}^n)$, and the following relation holds:

$$\widetilde{\mathcal{R}}_{\log M}(\mathrm{id}_{\mathbb{B}^n}, \mathcal{M}) \neq S^0(M, \mathbb{B}^n) \setminus (\mathrm{supp}\, S^0(\mathbb{B}^n) \cup \mathrm{ex}\, S^0(\mathbb{B}^n)).$$

A similar conclusion as in the above, may be established in the case of the family $\widetilde{\mathcal{R}}_{\log M}(\mathrm{id}_{\mathbb{B}^2}, \mathcal{N}_A)$, where $A \in \mathcal{A}$ is given by (2.2) (see [19]). More precisely, if $\lambda \in [1, 2)$, A is the matrix (2.2), F is given by (3.4), and $M = 2 \sup\{\| F(z)\| : z \in \mathbb{B}^2\}$, then (see [19, Remark 7.12])

$$\widetilde{\mathcal{R}}_{\log M}(\mathrm{id}_{\mathbb{B}^2}, \mathcal{N}_A) \subsetneqq S_A^0(M^\lambda, \mathbb{B}^2) \setminus (\operatorname{supp} S_A^0(\mathbb{B}^2) \cup \operatorname{ex} S_A^0(\mathbb{B}^2)).$$

Remark 4.12 The authors in [17, Theorem 4.6] obtained the sharp estimates for $\|f(z)\|$, $z \in \mathbb{B}^n$, in the case that $f \in \widetilde{\mathcal{R}}_{\log M}(\mathrm{id}_{\mathbb{B}^n}, \mathcal{N}_A)$ and $A \in \mathcal{A}$ is a diagonal matrix. In particular, if $A = I_n$ and $f \in \widetilde{\mathcal{R}}_{\log M}(\mathrm{id}_{\mathbb{B}^n}, \mathcal{M})$, then the following sharp estimates hold (see [16, Theorem 3.6]):

$$p_\pi^M(\|z\|) \leq \|f(z)\| \leq p_0^M(\|z\|), \quad z \in \mathbb{B}^n,$$

where $p_\alpha^M(\zeta) = Mk_\alpha^{-1}(\frac{1}{M}k_\alpha(\zeta))$, $k_\alpha(\zeta) = \frac{\zeta}{(1-e^{i\alpha}\zeta)^2}$, $\zeta \in \mathbb{U}$, is the rotation of the Koebe function, and $\alpha \in \mathbb{R}$ (see [34, Theorem 1], in the case $n = 1$).

Remark 4.13 Assume that $f \in \widetilde{\mathcal{R}}_{\log M}(\mathrm{id}_{\mathbb{B}^n}, \mathcal{M})$. Then $|V(P_2)| \leq 2\left(1 - \frac{1}{M}\right)$, where $P_2(z) = \frac{1}{2}D^2f(0)(z^2)$, $z \in \mathbb{C}^n$. This estimate is sharp, and equality holds for the mapping $F^M \in \widetilde{\mathcal{R}}_{\log M}(\mathrm{id}_{\mathbb{B}^n}, \mathcal{M})$ given by $F^M(z) = MF^{-1}(F(z)/M)$, $z \in \mathbb{B}^n$, where F is the Koebe mapping given by (3.5) (see [16, Corollary 3.13]).

Note that F^M is also a support point of $\widetilde{\mathcal{R}}_{\log M}(\mathrm{id}_{\mathbb{B}^n}, \mathcal{M})$ with respect to the linear functional $\Lambda : H(\mathbb{B}^n) \to \mathbb{R}$ given by $\Lambda(f) = \Re\langle\frac{1}{2}D^2f(0)(e_1^2), e_1\rangle, f \in H(\mathbb{B}^n)$, where $e_1 = (1, 0, \dots, 0)$ (cf. [16, Theorem 2.8]).

In connection with the above results, we formulate the following problems.

Problem 4.14 *Let $A \in \mathcal{A}$, $M > 1$, and $n \geq 2$. Find the sharp estimates for $\|f(z)\|$, $\|Df(z)\|$, and $|\det Df(z)|$, $z \in \mathbb{B}^n$, when $f \in \widetilde{\mathcal{R}}_{\log M}(\mathrm{id}_{\mathbb{B}^n}, \mathcal{N}_A)$. Also, find the sharp upper bounds for $|V(P_m)|$ and $\|P_m\|$, $m \geq 2$, where $P_m(z) = \frac{1}{m!}D^mf(0)(z^m)$, $z \in \mathbb{C}^n$.*

Problem 4.15 *(cf. [17]) Let $M \in (1, \infty)$, $A \in \mathcal{A}$, and let $n \geq 2$. Characterize all extreme points and support points of $\widetilde{\mathcal{R}}_{\log M}(\mathrm{id}_{\mathbb{B}^n}, \mathcal{N}_A)$.*

It is known that $\widetilde{\mathcal{R}}_{\log M}(\mathrm{id}_{\mathbb{U}}, \operatorname{ex} \mathcal{M}) = S(M)$ for $M > 1$, and $\widetilde{\mathcal{R}}_{\infty}(\mathrm{id}_{\mathbb{U}}, \operatorname{ex} \mathcal{M}) = S$ (see [11]; see also [34], [36, Theorem 1.47]). In higher dimensions, the following density result was recently obtained in [17, Theorem 4.14].

Theorem 4.16 *Let $A \in \mathcal{A}$ and let $M \in (1, \infty)$. Then*

$$\overline{\widetilde{\mathcal{R}}_{\log M}(\mathrm{id}_{\mathbb{B}^n}, \operatorname{ex} \mathcal{N}_A)} = \widetilde{\mathcal{R}}_{\log M}(\mathrm{id}_{\mathbb{B}^n}, \mathcal{N}_A).$$

The authors in [17, Conjecture 4.16] conjectured that if $n \geq 2$ and $A \in \mathcal{A}$, then $\overline{\widetilde{\mathcal{R}}_{\infty}(\mathrm{id}_{\mathbb{B}^n}, \operatorname{ex} \mathcal{N}_A)} = S_A^0(\mathbb{B}^n)$. In [25, Theorem 2] it was proved that the above conjecture is true.

Acknowledgements I. Graham was partially supported by the Natural Sciences and Engineering Research Council of Canada under Grant A9221. H. Hamada was partially supported by JSPS KAKENHI Grant Number JP16K05217.

References

1. Arosio, L.: Resonances in Loewner equations. Adv. Math. **227**, 1413–1435 (2011)
2. Arosio, L., Bracci, F., Hamada, H., Kohr, G.: An abstract approach to Loewner chains. J. Anal. Math. **119**, 89–114 (2013)
3. Arosio, L., Bracci, F., Wold, F.E.: Solving the Loewner PDE in complete hyperbolic starlike domains of $\mathbb{C}^n$. Adv. Math. **242**, 209–216 (2013)
4. Bracci, F.: Shearing process and an example of a bounded support function in $S^0(\mathbb{B}^2)$. Comput. Methods Funct. Theory **15**, 151–157 (2015)
5. Bracci, F., Roth, O.: Support points and the Bieberbach conjecture in higher dimension. Preprint (2016); arXiv: 1603.01532
6. Bracci, F., Contreras, M.D., Díaz-Madrigal, S.: Evolution families and the Loewner equation II: complex hyperbolic manifolds. Math. Ann. **344**, 947–962 (2009)
7. Bracci, F., Elin, M., Shoikhet, D.: Growth estimates for pseudo-dissipative holomorphic maps in Banach spaces. J. Nonlinear Convex Anal. **15**, 191–198 (2014)
8. Bracci, F., Graham, I., Hamada, H., Kohr, G.: Variation of Loewner chains, extreme and support points in the class S^0 in higher dimensions. Constr. Approx. **43**, 231–251 (2016)
9. Duren, P.: Univalent Functions. Springer, New York (1983)
10. Duren, P., Graham, I., Hamada, H., Kohr, G.: Solutions for the generalized Loewner differential equation in several complex variables. Math. Ann. **347**, 411–435 (2010)
11. Goodman, G.S.: Univalent functions and optimal control. Ph.D. Thesis, Stanford University (1968)
12. Graham, I., Kohr, G.: Geometric Function Theory in One and Higher Dimensions. Marcel Dekker, New York (2003)
13. Graham, I., Hamada, H., Kohr, G.: Parametric representation of univalent mappings in several complex variables. Can. J. Math. **54**, 324–351 (2002)
14. Graham, I., Kohr, G., Pfaltzgraff, J.A.: The general solution of the Loewner differential equation on the unit ball of $\mathbb{C}^n$. In: Complex Analysis and Dynamical Systems II. Contemporary Mathematics, vol. 382, pp. 191–203. American Mathematical Society, Providence (2005)
15. Graham, I., Hamada, H., Kohr, G., Kohr, M.: Asymptotically spirallike mappings in several complex variables. J. Anal. Math. **105**, 267–302 (2008)
16. Graham, I., Hamada, H., Kohr, G., Kohr, M.: Extreme points, support points and the Loewner variation in several complex variables. Sci. China Math. **55**, 1353–1366 (2012)
17. Graham, I., Hamada, H., Kohr, G., Kohr, M.: Extremal properties associated with univalent subordination chains in $\mathbb{C}^n$. Math. Ann. **359**, 61–99 (2014)
18. Graham, I., Hamada, H., Honda, T., Kohr, G., Shon, K.H.: Growth, distortion and coefficient bounds for Carathéodory families in $\mathbb{C}^n$ and complex Banach spaces. J. Math. Anal. Appl. **416**, 449–469 (2014)
19. Graham, I., Hamada, H., Kohr, G., Kohr, M.: Support points and extreme points for mappings with A-parametric representation in $\mathbb{C}^n$. J. Geom. Anal. **26**, 1560–1595 (2016)
20. Gurganus, K.: Φ-like holomorphic functions in $\mathbb{C}^n$ and Banach spaces. Trans. Am. Math. Soc. **205**, 389–406 (1975)
21. Gustafson, K.E., Rao, D.K.M.: Numerical Range. The Field of Values of Linear Operators and Matrices. Springer, New York (1997)
22. Hamada, H.: Polynomially bounded solutions to the Loewner differential equation in several complex variables. J. Math. Anal. Appl. **381**, 179–186 (2011)

23. Hamada, H.: Approximation properties on spirallike domains of $\mathbb{C}^n$. Adv. Math. **268**, 467–477 (2015)
24. Harris, L.: The numerical range of holomorphic functions in Banach spaces. Am. J. Math. **93**, 1005–1019 (1971)
25. Iancu, M.: A density result for parametric representations in several complex variables. Comput. Methods Funct. Theory **15**, 247–262 (2015)
26. Kirwan, W.E.: Extremal properties of slit conformal mappings. In: Brannan, D., Clunie, J. (eds.) Aspects of Contemporary Complex Analysis, pp. 439–449. Academic Press, London (1980)
27. Muir, J.: A Herglotz-type representation for vector-valued holomorphic mappings on the unit ball of $\mathbb{C}^n$. J. Math. Anal. Appl. **440**, 127–144 (2016)
28. Pell, R.: Support point functions and the Loewner variation. Pac. J. Math. **86**, 561–564 (1980)
29. Pfaltzgraff, J.A.: Subordination chains and univalence of holomorphic mappings in $\mathbb{C}^n$. Math. Ann. **210**, 55–68 (1974)
30. Pommerenke, C.: Univalent Functions. Vandenhoeck and Ruprecht, Göttingen (1975)
31. Poreda, T.: On the univalent holomorphic maps of the unit polydisc in $\mathbb{C}^n$ which have the parametric representation, I-the geometrical properties. Ann. Univ. Mariae Curie Skl. Sect. A. **41**, 105–113 (1987)
32. Poreda, T.: On the univalent holomorphic maps of the unit polydisc in $\mathbb{C}^n$ which have the parametric representation, II-the necessary conditions and the sufficient conditions. Ann. Univ. Mariae Curie Skl. Sect. A. **41**, 115–121 (1987)
33. Poreda, T.: On generalized differential equations in Banach spaces. Diss. Math. **310**, 1–50 (1991)
34. Prokhorov, D.V.: Bounded univalent functions. In: Kühnau R. (ed.) Handbook of Complex Analysis: Geometric Function Theory, vol. I, pp. 207–228. Elsevier, New York (2002)
35. Reich, S., Shoikhet, D.: Nonlinear Semigroups, Fixed Points, and Geometry of Domains in Banach Spaces. Imperial College Press, London (2005)
36. Roth, O.: Control theory in $\mathcal{H}(\mathbb{D})$. Dissertation, Bayerischen University Würzburg (1998)
37. Roth, O.: Pontryagin's maximum principle for the Loewner equation in higher dimensions. Can. J. Math. **67**, 942–960 (2015)
38. Schleissinger, S.: On support points of the class $S^0(\mathbb{B}^n)$. Proc. Am. Math. Soc. **142**, 3881–3887 (2014)
39. Suffridge, T.J.: Starlikeness, convexity and other geometric properties of holomorphic maps in higher dimensions. In: Lecture Notes in Mathematics, vol. 599, pp. 146–159. Springer, Berlin (1977)
40. Voda, M.: Loewner theory in several complex variables and related problems. PhD. Thesis, University of Toronto (2011)
41. Voda, M.: Solution of a Loewner chain equation in several complex variables. J. Math. Anal. Appl. **375**, 58–74 (2011)

Printed in the USA
By Bookmasters

Printed in the United States
By Bookmasters